Lecture Notes in Engineering

Edited by C. A. Brebbia and S. A. Orszag

5

Madassar Manzoor

Heat Flow Through Extended Surface Heat Exchangers

Springer-Verlag Berlin Heidelberg New York Tokyo 1984

Series Editors
C. A. Brebbia · S. A. Orszag

Consulting Editors
J. Argyris · K.-J. Bathe · A. S. Cakmak · J. Connor · R. McCrory
C. S. Desai · K.-P. Holz · F. A. Leckie · F. Pinder · A. R. S. Pont
J. H. Seinfeld · P. Silvester · P. Spanos · W. Wunderlich · S. Yip

Author
Madassar Manzoor
Dept. of Mathematical Sciences
University of Durham
Durham DH1 3LE
England

ISBN 3-540-13047-0 Springer-Verlag Berlin Heidelberg New York Tokyo
ISBN 0-387-13047-0 Springer-Verlag New York Heidelberg Berlin Tokyo

This work is subject to copyright. All rights are reserved, whether the whole or part of the material is concerned, specifically those of translation, reprinting, re-use of illustrations, broadcasting, reproduction by photocopying machine or similar means, and storage in data banks.
Under § 54 of the German Copyright Law where copies are made for other than private use, a fee is payable to 'Verwertungsgesellschaft Wort', Munich.

© Springer-Verlag Berlin Heidelberg 1984
Printed in Germany

Printing and binding: Beltz Offsetdruck, Hemsbach/Bergstr.
2061/3020-543210

To

Zahida

ABSTRACT

In conventional heat transfer equipment heat is exchanged between two fluids which are physically separated by a solid interface. Typical examples of such equipment can be found in devices as varied as air-conditioning units, aircraft and motor-vehicle engines, chemical processing plant, domestic refrigerators and radiators, electrical and electronic equipment, heat rejection systems in space vehicles, nuclear fuel storage tanks, and steam turbine installations. In these devices the separating interface invariably comprises a primary surface, such as a plane wall or a cylindrical tube, with slender metallic strips, referred to as extended surfaces, attached to one side. Consequently, such devices are classified as extended surface heat exchangers. The extended surfaces offer a convenient means of achieving a large heat transfer surface area without the use of excessive amounts of primary surface, and are mainly employed in situations in which there is a particular emphasis on minimising either the size or the weight of the heat transfer equipment.

The heat flow within extended surface heat exchangers is conventionally analysed on the basis of several simplifying assumptions. Recent investigations into the applicability of these assumptions have raised serious doubts concerning their validity. The primary objective of the work presented in this thesis has been to continue these investigations with a view to developing more representative models. This involved (a) a thorough revision of the existing methods of analysis, (b) the formulation of considerably more sophisticated models than previously examined, and (c) the development of appropriate analytical and numerical solution techniques. A most significant feature of this work has been the introduction of the BOUNDARY INTEGRAL EQUATION METHOD to the field of extended surface heat transfer. The advanced solution capabilities offered by this method are emphasised by the fact that much of the work presented in this thesis could not have been achieved by other comparable numerical techniques, such as the finite-difference and finite-element methods.

C O N T E N T S

Chapter 1   INTRODUCTION
1.1 Heat Exchangers   3
1.2 Design Parameters   4
1.3 Design Techniques   12
1.4 Limitations of the classical assumptions   14

Chapter 2   THE ONE-DIMENSIONAL ANALYSIS OF FIN ASSEMBLY HEAT TRANSFER
2.1 The one-dimensional analysis of fin assembly heat transfer   35

Chapter 3   THE TWO-DIMENSIONAL ANALYSIS OF FIN ASSEMBLY HEAT TRANSFER
3.1 The numerical solution of plane potential problems by improved
     boundary integral equation methods   61
3.2 The boundary integral equation analysis of transmission line
     singularities   89
3.3 The boundary integral equation analysis of fin assembly heat
     transfer   103
3.4 The analysis of fin assembly heat transfer by a series
     truncation method   130
3.5 The two-dimensional analysis of fin assembly heat transfer:
     A comparison of solution techniques   143

Chapter 4   THE ANALYSIS OF FIN RADIATION
4.1 The boundary integral equation analysis of non-linear plane
     potential problems   171
4.2 Improved formulations for the analysis of convecting and
     radiating finned surfaces   191

Chapter 5   THE APPLICABILITY OF THE PERFECT CONTACT ASSUMPTION
5.1 The effects of surface roughness on the performance of extended
     surface heat exchangers   231
5.2 The effects of interfacial bonding on the performance of
     extended surface heat exchangers   249

Chapter 6   CONCLUSIONS
6.1 Discussion and Conclusions   275

CHAPTER 1

INTRODUCTION

1.1 HEAT EXCHANGERS

Any apparatus which facilitates the exchange of heat between two fluids may be classified as a heat exchanger. The diversity of the applications in which heat exchanging apparatus are utilised covers an extensive range of equipment, varying in technological sophistication and size from domestic radiators and refrigerators, through aircraft and motor vehicle engines, to chemical processing plant. As a consequence, many different forms of heat exchangers have been developed. These are usually categorised as either <u>recuperators</u> or <u>regenerators</u> depending upon the process by which the heat exchange between the two heat transfer fluids is achieved. In recuperators the two heat transfer fluids simultaneously flow across the opposing surfaces of a solid interface, and the heat exchange occurs through this interface. A typical example is the radiator used in water cooled internal combustion engines; this device effects the transfer of heat from water circulating within its interior to air streaming across its exterior. Thus, recuperative heat exchangers facilitate a continuous exchange of heat. In contrast, in regenerators the heat transfer process is of a periodic nature. The heat exchange is achieved by alternatively passing the two heat transfer fluids over the same heat transfer surface; this surface absorbs heat from one fluid and subsequently releases this to the second fluid. Thus, regenerative heat exchangers are inevitably less effective than recuperators. However, they offer a convenient alternative for certain applications in which the use of recuperative heat exchangers is either impractical or unfeasible, e.g. in steelworks, the exhaust

gases from one furnace are used to preheat the blast air for a
second furnace using a rugged ceramic material, namely firebrick,
as a regenerative heat exchanger; the elevated temperature and
corrosive state of the exhaust gases precludes the application of
a recuperative heat exchanger.

In the vast majority of recuperative heat transfer equipment,
the recuperative interface which separates the two heat transfer
fluids is comprised of a <u>primary surface</u>, such as a plane wall or a
cylindrical tube, with slender metallic strips, referred to as
<u>extended surfaces</u>, attached to one side, e.g. as shown in Fig. 1.
Recuperators which utilise interfaces of this type are classified
as <u>extended surface heat exchangers</u> in order to distinguish them
from devices which rely solely on the primary interface to effect
the heat transfer. The extended surfaces, or fins as they are more
commonly known, offer a convenient and practical means of achieving
a large heat transfer surface, without the use of excessive amounts
of primary surface, and are mainly employed in situations where there
is a particular emphasis on minimising either the size or the weight
of the heat exchanger.

1.2 DESIGN PARAMETERS

The effective design of recuperative heat transfer equipment
requires an accurate prediction of the heat flow. This, in turn,
requires a thorough knowledge of the heat transfer characteristics
of both the recuperative interface and the two heat transfer fluids.
It has been established that the principal factors governing the heat
flow are the configuration and physical properties of the recuperative
interface, and the flow characteristics, physical properties and

operating temperatures of the two heat tranfer fluids, collectively referred to as the operating conditions, e.g. [1]. In practice, mainly as a consequence of the diversity of the applications in which recuperative heat exchangers are utilised, there are innumerable variants in all of these factors:

(a) Configuration of Recuperative Interface

It was explained earlier that in the vast majority of recuperative heat transfer equipment, the recuperative interface is comprised of a primary surface with extended surfaces appended to one side. The primary surface usually takes the form of either a plane wall or a cylindrical tube. However, the extended surfaces are not so limited in variety. They range from pin fins of cylindrical, conical and parabolic profile, through longitudinal fins of rectangular, trapezoidal and parabolic profile, to annular and helical fins of rectangular, trapezoidal and parabolic profile; some of the more frequently encountered configurations are shown in Fig. 1. The pin fins and the longitudinal fins are used with both plane walls and cylindrical tubes, but the annular and helical fins are only suitable for use with cylindrical tubes.

(b) Physical Properties of Recuperative Interface

The heat flow within the recuperative interface occurs by the process of thermal conduction, e.g. [2,3,4]. Consequently, the thermal conductivity of the interface is one of the most significant design parameters. This is emphasised by the fact that high conductivity metals such as aluminium and copper are extensively used for the fabrication of recuperative heat exchangers. However, these high conductivity metals are unsuitable for applications involving either severely corrosive fluid environments or extremely elevated operating temperatures. Consequently, heat exchangers fabricated from alloys such as brass and stainless-steel are also widely used. These alloys are more resistant to corrosion and can withstand higher operating temperatures than aluminium or copper, but suffer from the disadvantage that their thermal conductivities are substantially lower. In applications where corrosive conditions are evident in only one of the fluid environments, the combination of alloy primary surface and aluminium or copper extended surfaces provides the most effective heat transfer surface.

The heat flow within the recuperative interface is usually analised on the basis that the thermal conductivities of the various components of the recuperative interface are invariant, e.g. [1,5-10] This, in general, facilitates a considerable reduction in the complexity of the analysis. However, it is well known that variations in temperature can affect the thermal conductivity [2,3,4], e.g. as shown in Table 1:

Table 1. Variation of Thermal Conductivity with Temperature

Material	Thermal Conductivity (W/m K)			
	0 C	100 C	300 C	500 C
Aluminium	203	206	230	268
Brass	97	104	114	-
Copper	388	377	367	358
Nickel	60	59	55	-
Stainless-Steel	14	16	19	21

Nevertheless, in the vast majority of applications, the difference in the operating temperatures of the two heat transfer fluids is sufficiently restricted to permit the variation in the thermal conductivity to be neglected.

(c) Operating Conditions

From the preceeding discussion it is apparent that there is an extensive range of recuperative heat transfer equipment. This is a consequence of the wide variety of operating conditions under which recuperative heat exchangers are required to perform. The extent of this variety can best be illustrated by a detailed examination of the different types of heat transfer duties performed by recuperative heat exchangers:

(i) Heating

A heat exchanger which effects the transfer of heat from one fluid to another for the purpose of heating the cooler fluid is classified as a heater. Typical examples include the domestic radiator and the heating equipment used in tanks in which viscous fluids, such as crude oil, are stored. In the former, heat is transferred from heated water to atmospheric air, whilst in the latter, the viscous fluid is heated in order to prevent the formation of a sludge by the exchange of heat with a high-temperature heat transfer medium such as condensing steam.

(ii) Cooling

The cooling of a hot fluid by the rejection of its heat to a cooler fluid is by far the most popular of the functions performed by recuperative heat transfer equipment. Heat exchangers used for this purpose are generally referred to as coolers. One example of such a device is the radiator in water cooled internal combustion engines. As explained earlier, this device effects the cooling of heated water circulating within its interior by the dissipation of heat to air streaming across its exterior. Other examples include the pipes in air-conditioning plant and the oil-cooling apparatus employed in large diesel engines; in the air-conditioning equipment, heat is transferred from air to a refrigerating brine, whilst in the oil-cooler, heat is dissipated using water as the coolant.

(iii) Evaporating and Condensing

Recuperative heat exchangers which are employed in order to effect a change of phase of one of the heat transfer fluids are classified as either <u>evaporators</u> or <u>condensers</u>, depending upon whether the change of phase is from liquid to vapour or from vapour to liquid. The heat transfer components in gas cooled nuclear reactors, and boilers used in conjunction with furnaces in the metallurgical industries constitute examples of equipment used for evaporating. In the nuclear reactor, heat from hot carbon dioxide is used in order to effect the boiling of water, whilst in the boiler, the same function is performed by means of hot flue gases from the furnaces. At the opposite extreme are devices, such as the heat transfer equipment in steam turbines and gas liquefaction plant, which effect a vapour to liquid phase change; the heat transfer equipment in steam turbine installations uses water in order to cool and condense steam, whilst the gas liquefaction apparatus performs its function by the extraction of heat using some form of refrigerant.

It is important to note that a heat exchanger is classified as an evaporator or a condenser only if its <u>primary</u> function is to effect a change of phase of one of the heat transfer fluids. There are many heat exchangers which are not categorised as evaporators or condensers even though their operation involves the change of phase of one of the heat transfer fluids, e.g. the domestic refrigerator is classified as a cooler although its operation involves the evaporation of the refrigerant Freon, and the heat transfer equipment used in large oil storage tanks is categorised as a heater regardless of the fact that the heating is achieved by means of condensing steam. In such devices, the change of phase is not itself the primary objective of the heat transfer process, but occurs in order to effect some other heat transfer process, usually the heating or cooling of the associated heat transfer fluid.

All the examples cited in the preceeding discussion have been restricted to applications in which heat transfer fluids are present on both sides of the recuperative interface. A further range of heat transfer equipment is introduced for applications in which one of the surfaces of the recuperative interface operates in a vacuum, e.g. the cooling equipment used in space vehicles dissipates waste heat into free space. This class of heat exchanger is generally categorised as a <u>radiator</u> because the surface heat dissipation occurs solely by means of thermal radiation; in the presence of a heat transfer fluid the surface heat transfer occurs by a combination of convection and radiation, and therefore, devices such as the domestic radiator and the radiator in motor vehicle engines cannot be classified as "true" radiators.

The theoretical analysis of the heat flow within a recuperative heat exchanger necessitates a mathematical description of the surface heat transfer. As explained earlier, the surface heat transfer occurs by a combination of convection and radiation. The convective heat transfer is a complex combination of conductive heat exchange and fluid motion, and as such, its description requires a thorough understanding of the principles of thermal conduction and fluid dynamics [2,3,4]. In addition, an extensive knowledge of the principles of boundary layer theory is also required because the process of convection between a surface and a fluid is intimately concerned with the heat conduction and mass transfer in the fluid layers in the immediate vicinity of the surface. The combined complexity of these features precludes an exact mathematical representation. Consequently, in present design techniques, the convective heat transfer process is described employing Newton's Law of Cooling [2,3,4] which stipulates that the convective heat flow from a surface to a fluid is directly proportional to the area of the surface exposed to the fluid, and the difference between the temperature of the surface and that of the fluid. This relationship is usually expressed in the form,

$$dQ(p) = h(p) \, dA(p) \, (T(p) - T_\infty)$$

where h is the constant of proportionality, generally referred to as the heat transfer coefficient. This is a gross quantity which represents the overall effect of the convective heat transfer process, but makes no attempt to describe the actual mechanics of this process. The numerical value of this parameter is governed, in particular, by the geometry of the surface, the velocity distribution within the fluid, the physical properties of the fluid and the temperature difference between the surface and the fluid [2,3,4]. Consequently, the heat transfer coefficient is unlikely to be invariant over a surface. Nevertheless, calculations are usually made on the basis of averaged values of the heat transfer coefficient because this, in general, facilitates a considerable reduction in the complexity of the associated analysis.

The process of radiative heat transfer differs in nature from both conduction and convection in that it occurs without the aid of a physical transport medium, by means of electromagnetic emission of thermal energy [11,12]. The rate at which radiant energy is dissipated from a surface is dependent upon the temperature of the surface, the substance of which the surface is composed and the nature of the surface structure, and is described by the Stefan-Boltzmann Law [11,12],

$$dQ(p) = dA(p) \, \epsilon\sigma \, T^4(p)$$

where ϵ, the emissivity of the surface, is a physical property which characterises the nature of the surface structure, and σ is the Stefan-Boltzmann constant which takes the value 5.67×10^{-8} W/m^2. If radiant energy from either the surrounding environment (e.g. free space) or adjacent surfaces (or both) impinges upon a surface then the net radiative heat loss from the surface is defined by the relation,

$$dQ(p) = dA(p) \, (\epsilon\sigma \, T^4(p) - (1-\alpha)G(p))$$

where α, the absorbtivity of the surface, is a physical property which is dependent upon the nature of the surface structure and denotes the fraction of the incident radiant energy absorbed by the surface and G denotes the amount of radiant energy incident upon the surface. It is important to note that this relation defines the total radiative heat loss from a surface. This includes both the radiant energy emitted by the surface solely by virtue of its temperature and also the reflected incident radiation.

The radiative surface heat transfer is usually neglected in design calculations because it introduces considerable complexity into the analysis by virtue of its non-linear nature. However, when radiative heat transfer is included, the analysis invariably involves idealisations as to the nature of the radiant energy. The early investigations restricted attention to black-body radiation for which the simplifications $\alpha = 1$ and $\varepsilon = 1$ apply. The most recent investigations have considered gray-body radiation which permits α and ε to take more realistic values, but still requires that α be identically equal to ε.

1.3 DESIGN TECHNIQUES

The earliest mathematically rigorous investigation pertaining to the heat flow within extended surface heat exchangers was performed by Harper and Brown [1] in 1922. This study presented a simple theoretical representation of the heat flow within a longitudinal rectangular fin attached to an isothermal plane surface, as shown schematically in Fig. 2. This theoretical representation was formulated on the basis of several simplifying assumptions. In particular,

i) the thermal conductivity of the fin is invariant,
ii) the surface heat dissipation is purely convective and the associated heat transfer coefficient is uniform over the entire heat exchanging surface of the fin,
iii) the transverse (i.e. y-direction, Fig. 1) variation of the temperature distribution within the fin is negligible, and
iv) there is perfect contact between the fin and the supporting surface.

These simplifications, which are now commonly referred to as the classical or conventional assumptions, enabled the derivation of closed form solutions for both the temperature distribution and the heat transfer rate of the fin. In order to present these solutions in a form convenient for design purposes a dimensionless quantity referred to as the fin efficiency was introduced. This quantity was defined as the ratio of the rate at which heat is dissipated from the fin to the rate at which heat would be dissipated if the entire fin was maintained at the same temperature as the supporting surface. It had the inherent advantage that it could be parameterised by a single variable comprised of a dimensionless combination of the fin length, fin thickness, fin thermal conductivity and surface heat transfer coefficient. Thus, the fin performance could be characterised by a single curve, Fig. 3, which accounted for all possible variations in the system parameters.

The model upon which Harper and Brown [1] based their formulation is not representative of the conditions within an extended surface heat exchanger because if fails to account for either the convective heat exchange at the plain (i.e. unfinned) side of the primary interface or the conductive heat flow within the primary interface. Fortunately, their formulation is perfectly compatible with a technique, commonly referred to as the sum of Resistances method [4], which accounts for these factors and thereby enables the prediction of the performance of a fin assembly, such as that shown in Fig. 4. However, this technique is formulated, without a formal mathematical analysis, by analogy with electric circuit theory. It is shown in this study (Chapter 2 [13]) that this technique in fact has a mathematically rigorous foundation based upon the assumption that the heat flow within the recuperative interface is one-dimensional.

A very large proportion of the subsequent literature on the subject of extended surface heat transfer, e.g. [5-10], has adhered rigidly to the concepts introduced by Harper and Brown [1]. Consequently, there is now an extensive range of efficiency charts covering a wide variety of fin profiles in both the plane and annular geometries. As a result, present design techniques are based almost entirely upon this approach. However, recent investigations into the applicability of the underlying assumptions have raised serious doubts about their validity. The main aim of the work presented in this thesis has been to continue these investigations with a view to developing more representative models.

1.4 LIMITATIONS OF THE CLASSICAL ASSUMPTIONS

In this section the applicability of the classical assumptions is discussed in detail, and the improvements devised in this study are contrasted with the existing formulations.

(a) Non-Linear Thermal Conductivity

It was explained earlier that for the materials from which extended surface heat exchangers are commonly fabricated, the thermal conductivity is effectively constant under practical operating conditions. Nevertheless, investigations have been performed allowing for non-linear variation in thermal conductivity, e.g. [14-17]. These have confirmed that the assumption that the thermal condutivity is invariant is accurate provided that the temperature variations are not too excessive.

(b) Non-Uniform Heat Transfer

From the complex nature of convective heat transfer it is apparent that the surface heat transfer coefficient is unlikely to be invariant over the fin surface. In fact, numerous experimental investigations have shown that the distribution of the heat transfer coefficient is far from uniform, e.g. [18-21]. However, these investigations have failed to establish any generally applicable data; this is not surprising in view of the numerous features which can affect the convective heat transfer process, namely, the configuration and thermal conductivity of the recuperative interface, and the flow characteristics and thermal conductivity of the heat transfer fluid. However, there have also been several theoretical investigations into the effects of non-uniform heat transfer, e.g. [22,23,24]. These investigations have attempted to account for the non-uniformity by the use of the assumed distributions, e.g. Hans and Lefkowitz [22] considered heat transfer coefficients which varied as given powers of the displacement from the fin base, whilst Heggs et al [24] considered heat transfer coefficients which increased linearly from the fin base to the fin tip. These investigations showed that, in some cases, the assumed variations can give results which are in excellent agreement with those obtained experimentally. Unfortunately, for any given problem the appropriate variation cannot be prescribed and consequently uniform heat transfer coefficients are still extensively used.

(c) Two-Dimensional Effects.

The first investigation into the applicability of the one-dimensional approximation was performed by Harper and Brown [1] in their original study of 1922. They analysed the heat flow within the fin on the basis of two-dimensional heat flow and concluded that the one-dimensional approximation is accurate provided the fin length is very much larger than the fin thickness. However, in 1968 Irey [25] showed that the applicability of the one-dimensional approximation was dependent upon the transverse Biot number (Bi) and not the ratio of the fin length to the fin thickness. This result was subsequently verified by Levitsky [26] and Lau and Tan [27].

In order to investigate whether these findings also applied in the case of a fin assembly, Sparrow and Hennecke [28] considered two-dimensional heat flow in a system comprised of a longitudinal rectangular fin attached to a plane wall of infinite thickness. They discovered significant two-dimensional variations in the temperature distribution within the wall. These were attributed to the fact that appreciably more heat is channelled through the fin than through the unfinned surface adjacent to the fin. Sparrow and Lee [29], Suryanarayana [30] and Heggs and Stones [31] have reached the same conclusions for the case of fins attached to a wall of finite thickness, (i.e. for fin assemblies). In fact, Suryanarayana [30] reported that the solutions predicted by the one and two-dimensional formulations can differ by as much as 80 per cent. Thus, it was established that an accurate representation of the heat flow could only be achieved by considering both the wall and the fins simultaneously, and employing a two-dimensional analysis.

The steady-state two-dimensional analysis of the heat flow within a fin assembly comprised of longitudinal rectangular fins attached to a plane wall requires the solution of a plane Laplacian mixed boundary-value problem. In the published literature, solutions have been computed employing either the finite-difference (FD) or the finite-element (FE) methods. However, a comparison of

these two methods, performed by Stones [32], has revealed that for
a particular range of the system parameters neither method provides
satisfactory solutions. Consequently, in this study the performance
of two alternative solution techniques, namely the boundary integral
equation (BIE) method and the series truncation (ST) method, have
been investigated. Several different BIE implementations are
discussed in sections 3.1 [33], 3.2 [34], and then these are applied
to the fin assembly problem in section 3.3 [35]. The ST method is
described in section 3.4 [36]. Finally, a comparison of all four
solution techniques is presented in section 3.5 [37]. This
comparison indicates that the BIE method is by far the most
computationally efficient method for analysing fin assembly heat
transfer problems. Consequently, this method is used throughout
the subsequent work.

(d) Radiative Heat Dissipation

Radiation heat transfer is usually neglected in design
calculations e.g. [5-10]. However, it plays a significant role in
heat exchangers operating at high temperatures or in atmosphere-free
environments [2,3,4]. The original investigations into this aspect
considered isolated black fins operating in the absence of any
surface convection, e.g. [38,39]. However, there has been considerable
progress in this field. The most recent investigations [40,41,43]
have considered combined convective and radiative heat dissipation
from gray fins which radiatively interact with adjacent fins and also
the base-surface. These studies have shown that radiative heat
transfer must be accounted for except in situations in which the
convective heat transfer is extremely high. However, even these
advanced formulations are unsuitable for design purposes because they

are based upon one-dimensional analyses with attention restricted solely to the fin side.

For fin assembly problems involving radiative heat dissipation, a two-dimensional analysis of the heat flow requires the solution of a Laplacian mixed boundary-value problem involving non-linear boundary conditions. The solution to such a problem may be attempted by either the FD, FE or BIE methods; the ST method cannot handle the non-linear boundary conditions. Since the FD and FE methods display problems even in the linear case [32], it would be inappropriate to attempt to apply these methods. Thus, the BIE method appears to be the best alternative. However, prior to the present investigation, the application of the BIE method had been restricted entirely to linear problems e.g. [43-50]. Thus, in order to apply the BIE method it is first necessary to extend the BIE formulation in order to enable it to handle the non-linear boundary conditions. The techniques by which this extension can be achieved are explained in section 4.1 [51]. Then, in section 4.2 [52], the fin assembly problem is analysed for the case of combined convective and radiative surface heat dissipation. The general situation in which the fins and the base-surface have different thermal conductivities and different surface emissivities is examined, and all relevant radiant interactions are accounted for.

(e) Contact Resistance

The assumption that there is perfect contact between the fins and the supporting surface facilitates a considerable reduction in the complexity of the analysis. However, this assumption is strictly true only if the fins form an integral part of the base-surface. If the fins are bonded to the base surface by means of brazing, soldering or welding, then either the roughness of the contacting surfaces or the presence of the bonding material (or both) may prevent perfect contact. However, prior to the present study, there have only been three published articles in which the validity of the perfect contact assumption has been investigated [53,54,55]. All of these articles considered the case of spiral fins wound onto a cylindrical tube and examined the heat flow in the situation when differential expansion completely relaxes the contact between the fins and the tube. Since, in general, the bonding maintains contact, at least at discrete zones along the contact interface, the results of these investigations are of limited value.

In this study the effects of bonding on the heat flow through a fin assembly are investigated. In section 5.1 [56] a theoretical representation of surface roughness is devised in order to account for contacts in which the bonding material is only applied to the sides of the fin, e.g. as shown in Fig. 5. Then, in section 5.2 [57], the effects of introducing the bonding material between the fins and the supporting surface, as shown in Fig. 6, is investigated. In both cases the analysis is performed on the basis of two-dimensional heat flow and solutions are computed employing the BIE method; of the four aforementioned solution techniques, the BIE method most easily handles the geometrical complexities arising from the modelling of the contact interface.

NOMENCLATURE

Bi = ht/k, Biot number

dA incremental surface area, m^2

dQ incremental heat transfer rate, W

G irradiation, W/m^2

h heat transfer coefficient, W/mK

k thermal conductivity, $W/m^2 K$

ℓ fin length, m

p point on the heat transfer surface

t fin thickness, m

T temperature, K

w wall thickness, m

α absorbtivity

ε emissivity

σ Stefan-Boltzmann constant, $W/m^2 K^4$

η fin efficiency

REFERENCES

1. D.R. Harper and W.B. Brown, "Mathematical equations for heat conduction in the fins of air-cooled engines", National Advisory Committee for Aeronautics, Report 158, 1922.

2. D.Q. Kern and A.D. Kraus, Extended surface heat transfer, McGraw-Hill, New York, 1972.

3. A.J. Chapman, Heat transfer, Macmillan, New York, 1974.

4. F. Kreith, Principles of heat transfer, Harper and Row, New York, 1976.

5. W.G. Murray, "Heat dissipation through an aunnular disk or fin of uniform thickness", Journal of Applied Mechanics, Vol. 60, pp. A78-A80, 1938.

6. K.A. Gardner, "Efficiency of extended surface", Transactions of the ASME, Vol. 67, pp. 621-631, 1945.

7. A. Brown, "Optimum dimensions of uniform annular fins", International Journal of Heat and Mass Transfer, Vol. 8, pp. 655-662, 1965.

8. P.J. Smith and J. Sucec, "Efficiency of circular fins of triangular profile", Journal of Heat Transfer, Vo. 91, pp. 181-182. 1969.

9. S. Guceri and C.J. Maday, "A least weight circular cooling fin", Journal of Engineering for Industry, Vol. 97, pp. 1190-1193, 1975.

10. I. Mikk, "Convective fin of minimum mass", International Journal of Heat and Mass Transfer, Vol. 23, pp. 707-711, 1981.

11. E.M. Sparrow and R.D. Cess, Radiation heat transfer, Brooks-Cole Publishing Company, Belmont, California, 1970.

12. R. Siegel and J.R. Howell, Thermal radiation heat transfer, McGraw-Hill, New York, 1972.

13. P.J. Heggs, D.B. Ingham and M. Manzoor, "The one-dimensional analysis of fin assembly heat transfer", submitted to Journal of Heat Transfer, 1981.

14. A. Aziz and S.M. Enamul-Huq, "Perturbation solution for convecting fin with variable thermal conductivity", Journal of Heat Transfer, Vol. 97, pp. 300-301, 1975.

15. A. Muzzio, "Approximate solution for convective fins with variable thermal conductivity", Journal of Heat Transfer, Vol. 98, pp. 680-682, 1976.

16. P. Razelos and K. Imre, "The optimum dimensions of circular fins with variable thermal parameters", Journal of Heat Transfer, Vol. 102, pp. 420-425, 1980.

17. A. Aziz and T.Y. Na, "Periodic heat transfer in fins with variable thermal parameters", International Journal of Heat and Mass Transfer, Vol. 24, pp. 1397-1404, 1981.

18. P.W. Wong, "Mass and heat transfer from circular finned cylinders", Journal of the Institution of Heating and Ventilating Engineers, Vol. 23, pp. 1-23, 1963.

19. J.W. Stachiewicz, "Effect of variation of local film coefficient on fin performance", Journal of Heat Transfer, Vol. 91, pp. 21-26, 1969.

20. V.F. Yudin and L.C. Tokhtorova, "Investigation of the correction factor ψ for the theoretical effectiveness of a round fin", Thermal Engineering, Vol. 20, pp. 66-68, 1973.

21. P.J. Heggs and P.R. Stones, "Improved design method for finned tube heat exchangers", Transactions of the Institution of Chemical Engineers, Vol. 58, pp. 147-154, 1980.

22. L.S. Han and S.G. Lefkowitz, ASME Paper 60-WA-41, 1960.

23. P.G. Barnett, "The influence of wall thickness, thermal conductivity and method of heat input on the heat transfer performance of some ribbed surfaces", International Journal of Heat and Mass Transfer, 1972.

24. P.J. Heggs, D.B. Ingham and M. Manzoor, "The effects of non-uniform heat transfer from an annular fin of triangular profile", Journal of Heat Transfer, Vol. 103, pp. 184-185, 1981.

25. R.K. Irey, "Errors in the one-dimensional fin solution", Journal of Heat Transfer, Vol. 90, pp. 175-176, 1968.

26. M. Levitsky, "The criterion for validity of the fin approximation", International Journal of Heat and Mass Transfer, Vol. 15, pp. 1960-1963, 1972.

27. W. Lau and C.W. Tan, "Errors in one-dimensional heat transfer analysis in straight and annular fins", Journal of Heat Transfer, Vol. 95, pp. 549-551, 1973.

28. E.M. Sparrow and D.K. Hennecke, "Temperature depression at the base of a fin", Journal of Heat Transfer, Vol. 92, pp. 204-206, 1970.

29. E.M. Sparrow and L. Lee, "Effects of fin-base temperature depression in a multifin array", Journal of Heat Transfer, Vol. 97, pp. 463-465, 1975.

30. N.V. Suryanarayana, "Two-dimensional effects on heat transfer from an array of straight fins", Journal of Heat Transfer, Vol. 99, p.p. 129-132, 1977.

31. P.J. Heggs and P.R. Stones, "The effects of dimensions on the heat flowrate through extended surfaces", Journal of Heat Transfer, Vol. 102, pp. 180-182, 1980.

32. P.R. Stones, Ph.D. Thesis, University of Leeds, 1980.

33. D.B. Ingham, P.J. Heggs and M. Manzoor, "The numerical solution of plane potential problems by improved boundary integral equation methods", Journal of Computational Physics, Vol. 42, pp 77-98, 1981.

34. D.B. Ingham, P.J. Heggs and M. Manzoor, "Boundary integral equation analysis of transmission line singularities", IEEE Transactions on Microwave Theory and Techniques, Vol. 29, pp. 1240-1243, 1981.

35. P.J. Heggs, D.B. Ingham and M. Manzoor, "Boundary integral equation analysis of fin assembly heat transfer", to appear in Numerical Heat Transfer.

36. P.J. Heggs, D.B. Ingham and M. Manzoor, "The analysis of fin assembly heat transfer by a series truncation method", to appear in Journal of Heat Transfer.

37. D.B. Ingham, P.J. Heggs and M. Manzoor, "The two-dimensional analysis of fin assembly heat transfer: A comparison of solution techniques", to appear in the Proceedings of the Second National Symposium on Numerical Methods in Heat Transfer, Hemisphere, Washington DC, 1982.

38. R.L. Chambers and E.V. Somers, "Radiation fin efficiency for one-dimensional heat flow in a circumar fin", Journal of Heat Transfer, Vol. 81, pp. 327-329, 1959.

39. J.G. Bartas and W.H. Sellers, "Radiation fin effectiveness", Journal of Heat Transfer, Vol. 82, pp. 73-75, 1960.

40. R.C. Donovan and W.M. Rohrer, "Radiative and convecting fins on a plane wall including mutual irradiation", Journal of Heat Transfer, Vol. 93, pp. 41-46, 1971.

41. M.N. Schnurr, "Radiation from an array of longitudinal fins of triangular profile", AIAA Journal, Vol. 13, pp. 691-693, 1975.

42. R.G. Eslinger and B.T.F. Chung, "Periodic heat transfer in radiating and convecting fins or fin arrays", AIAA Journal, Vol. 17, pp. 1134-1140, 1979.

43. G.T. Symm, "Treatment of singularities in the solution of Laplace's equation by an integral equation method", National Physical Laboratory, Report NAC31, 1973.

44. W.A. Bell, W.L. Meyer and B.T. Zinn, "Predicting the acoustics of arbitrarily shaped bodies using an integral equation approach", AIAA Journal, Vol. 15, pp. 813-820, 1977.

45. Y.S. Wu, F.J. Rizzo, D.J. Shippy and J.A. Wagner, "An advanced boundary integral equation method for two-dimensional electromagnetic field problems", Electric Machines and Electromechanics, Vol. 1, pp 303-313, 1977.

46. D.L. Clements and F.J. Rizzo, "A method for the numerical solution of boundary value problems governed by second-order elliptic equations", Journal of the Institute of Mathematics and its Applications, Vol. 22, pp. 197-202, 1978.

47. E. Alarcon, C. Brebbia and J. Dominguez, "The boundary element method in elasticity", International Journal of Mechanical Science, Vol. 20, pp. 625-639, 1978.

48. C.A.Brabbia, Boundary element methods for engineering, Pentech Press, London, 1980.

49. C.A.Brebbia and S.Walker, Boundary element techniques in engineering, Butterworths, London, 1980.

50. C.A.Brebbia, Progress in boundary element methods, Volume 1, Pentech Press, London, 1981.

51. D.B.Ingham, P.J.Heggs and M.Manzoor, "The boundary integral equation solution of non-linear plane potential problems", to appear in the Institute of Mathematics and Its Aplications Journal of Numerical Analysis.

52. D.B.Ingham, P.J.Heggs and M.Manzoor, "Improved formulations for the analysis of convecting and radiating finned surfaces", to appear in AIAA Journal.

53. K.A.Gardner and T.C.Carnovos, "Thermal-contact resistance in finned tubing", Journal of Heat Transfer, Vol.82, pp.279-284, 1960.

54. E.H.Young and D.E.Briggs, "Bond resistance of bimetallic finned tubes", Chemical Engineering Progress, Vol. 61, pp.71-76, 1965.

55. M.V.Kulkarni and E.H.Young, "Bimetallic finned tubes", Chemical Engineering Progress, Vol.62, pp.69-74, 1966.

56. P.J.Heggs, D.B.Ingham and M.Manzoor, "The effects of surface roughness on the performance of extended surface heat exchangers", submitted to Journal of Heat Transfer, 1981.

57. P.J.Heggs, D.B.Ingham and M.Manzoor, "The effects of interfacial bonding on the performance of extended surface heat exchangers", submitted to Journal of Engineering for Industry, 1981.

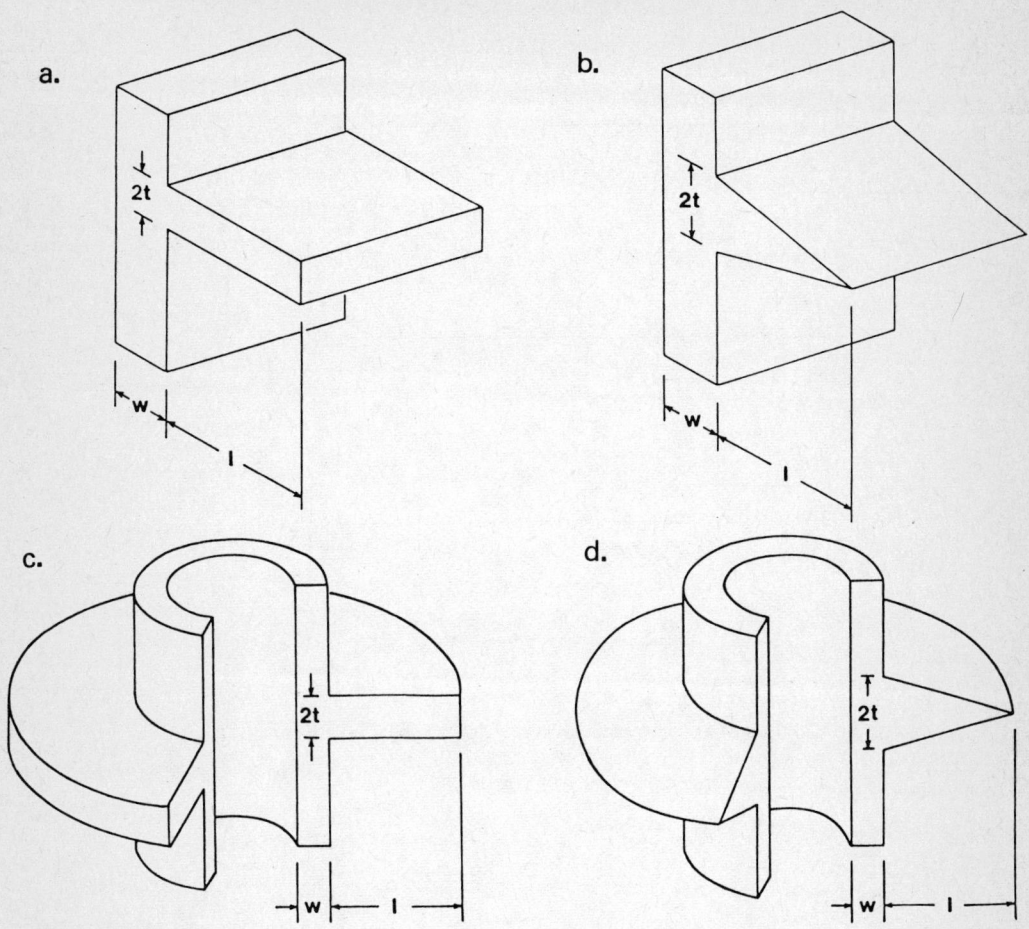

Fig. 1 Examples of some typical fin configurations

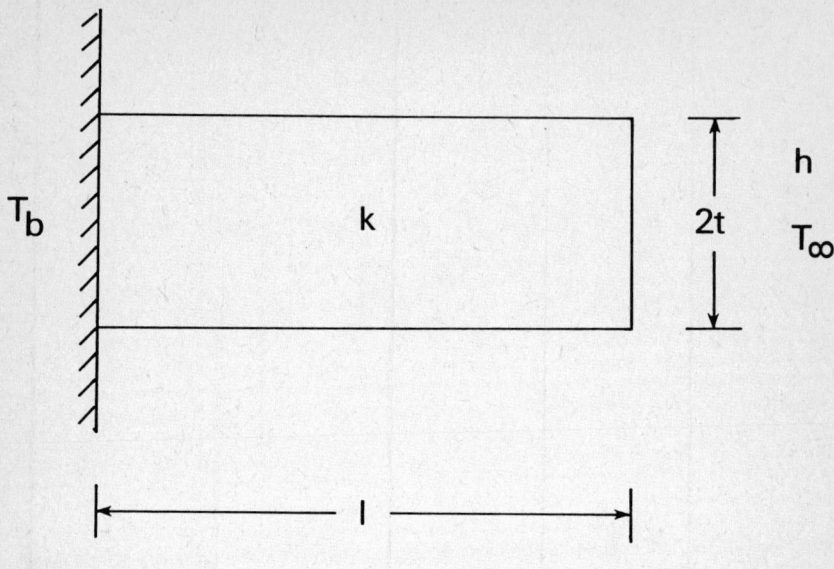

Fig. 2 Fin configuration investigated by Harper and Brown [1]

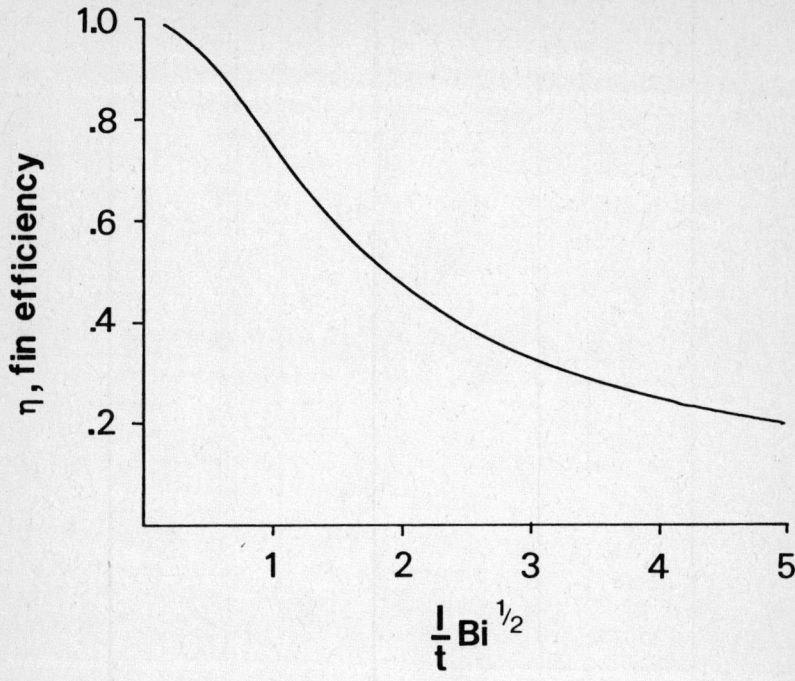

Fig. 3 The fin-efficiency of a longitudinal rectangular fin

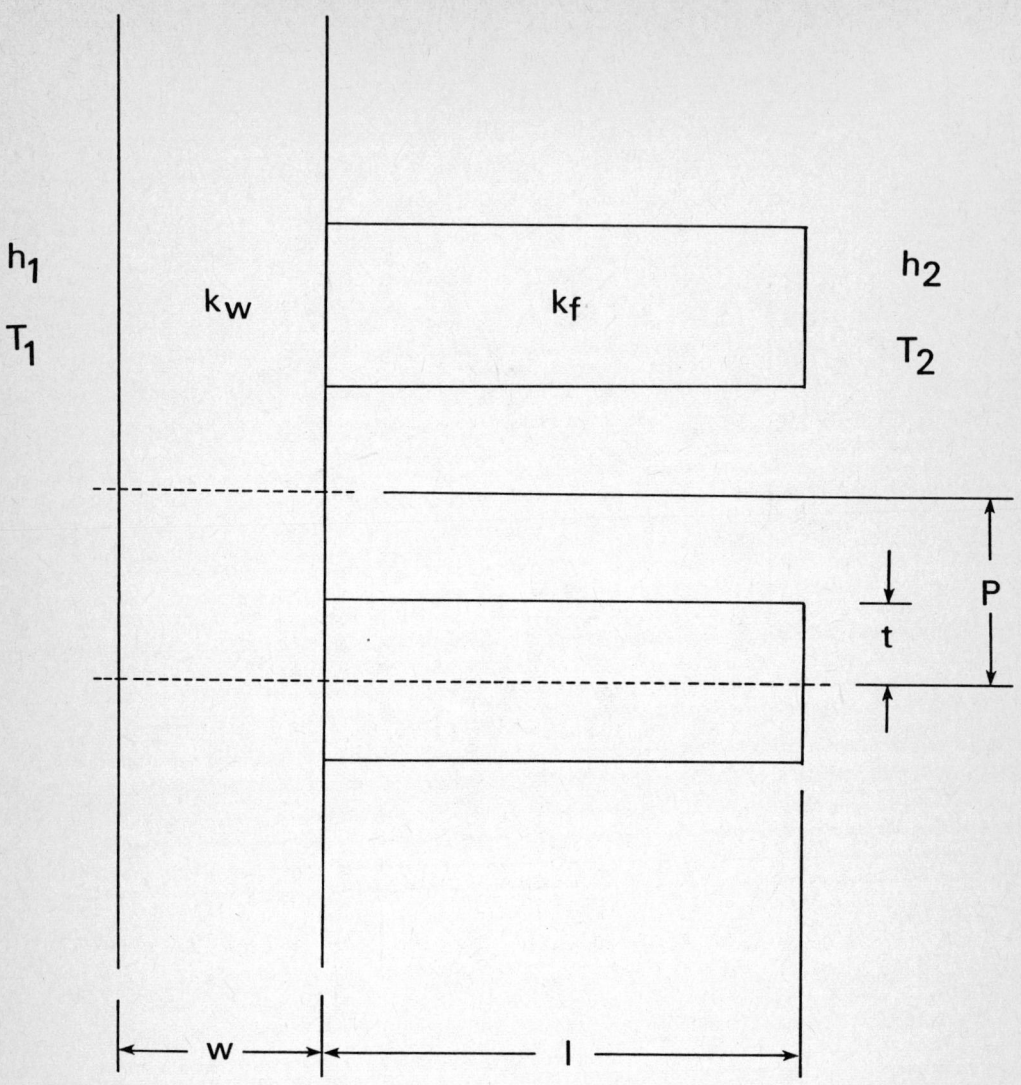

Fig. 4 Schematic representation of a fin assembly

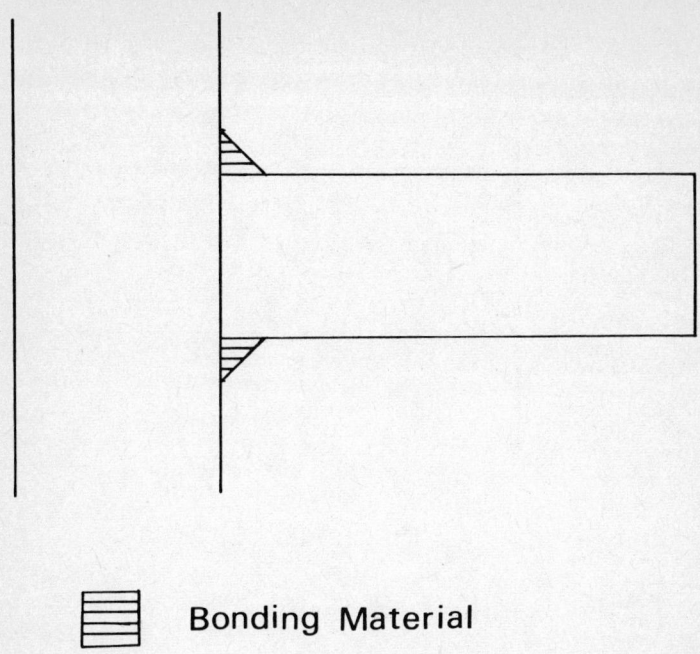

Fig. 5 Schematic representation of "exterior" bonding

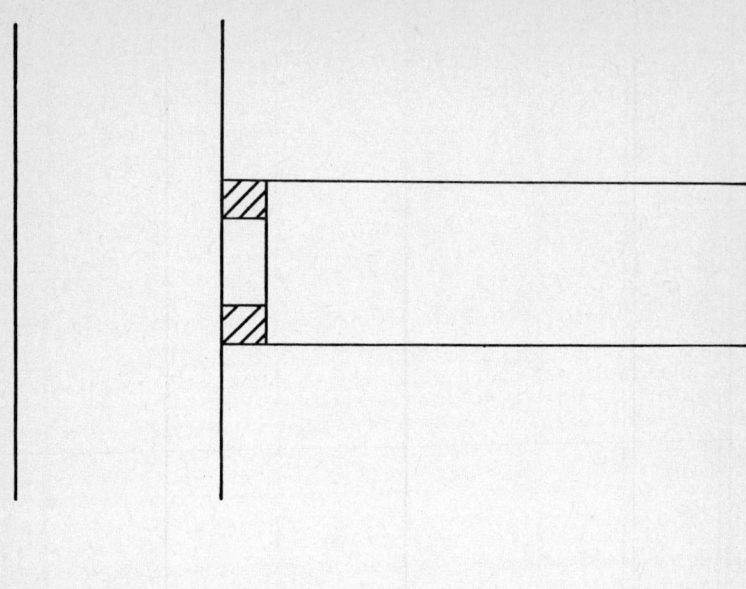

Fig. 6 Schematic representation of "interfacial" bonding

CHAPTER 2

THE ONE-DIMENSIONAL ANALYSIS OF FIN ASSEMBLY HEAT TRANSFER

2.1 THE ONE-DIMENSIONAL ANALYSIS OF FIN ASSEMBLY HEAT TRANSFER

ABSTRACT

The design of finned surfaces is conventionally performed in two stages. First, the fin efficiency is determined by simultaneously analysing the conductive heat flow within the fin, and the convective heat dissipation from the surface of the fin. Then, the effects of the thermal interaction between the supporting interface and the fins, and the convective heat exchange at the plain side of the supporting interface are incorporated by employing a technique based on electric circuit theory. In this study, it is shown that this technique in fact has a mathematically rigorous foundation. It is also shown that, for design purposes, there is a far superior alternative to the fin efficiency.

INTRODUCTION

It has long been recognized that the use of fins can facilitate an augmentation of the heat exchange between two fluids which are separated by a solid interface. The fins offer a convenient and practical means of achieving a large heat transfer surface without the use of excessive amounts of primary surface and are therefore widely used in situations where there is a particular emphasis on minimising either the size or the weight of the heat transfer equipment. These requirements place a considerable significance on the theoretical analysis of the heat flow because an accurate prediction of the thermal performance is essential for effective design. However, present design techniques are based upon a method of analysis which is far from rigorous. This method originated from the fundamentally innovative work of Harper and

Brown [1] and involves a process of analysis in which the fins and the supporting interface are effectively treated as completely separate entities. The heat flow within the fins is analysed solely by examining the heat conduction within the fins and the heat convection from the surface of the fins, e.g. [1-7]. The effects of such significant features as the heat conduction within the supporting interface, the proximity of adjacent fins, and the heat exchange from the plain (i.e. unfinned) side of the supporting interface are given no consideration whatsoever in the fin analysis, but are accounted for in a separate analysis. However, this secondary analysis, which is commonly referred to as the sum of resistances method, is not of a mathematically formal nature, but is performed by analogy with electric circuit theory [6].

The principal advantage of examining the fins in isolation from the supporting interface is that this permits a very concise non-dimensional representation of the fin heat transfer rate, commonly referred to as the fin efficiency [1-7]. The conciseness of this representation results from the fact that the fin efficiency can usually be parameterised by a relatively small number of bulk parameters instead of the actual fin parameters [1-7]. For example, the fin efficiency of a longitudinal fin of recangular profile can be parameterised by a single variable comprised of a non-dimensional combination of the fin length, fin thickness, fin thermal conductivity and surface heat transfer coefficient, and this enables the fin performance to be characterised by a single curve which accounts for all possible variations in the four fin parameters, e.g. [2]. As a consequence of this type of concise graphical presentation of design data, the fin efficiency is widely regarded as the ideal design tool. However, it conciseness in fact negates its value for design purposes because the effects of variations in the individual design parameters on the fin performance cannot easily be deduced from the fin efficiency graphs. Furthermore, since the fin efficiency is completely independent of both the heat conduction within the supporting interface, and the heat convection at the plain side of this interface, it gives practically no indication as to the effect of the presence of the fins, on the overall heat transfer.

In this study a rigorous theoretical method is devised for analysing the heat flow within finned surfaces. In contrast to the conventional two-stage method of analysis, this method is formulated by simultaneously examining the heat flow within both the fins and the supporting interface, i.e. by employing a fin assembly representation. The general situation in which the fins and the supporting interface have different thermal conductivities is considered, and the effects of the thermal interaction between the fins and the supporting interface are accounted for in a mathematically formal manner. A salient feature of this method is that it results in exactly the same expression for the overall heat transfer rate as that predicted by the sum of resistance approach, thus indicating that the present work is consistent with the conventional method of analysis. The most significant feature of the present work is the introduction of a superior alternative to the fin efficiency, namely the enhancement factor. This enhancement factor is not as concise as the fin efficiency, in that, it requires more charts in order to decribe the fin side heat transfer. However, these charts clearly show the effects of variations in the fin dimensions, interfin spacing, fin thermal conductivity and surface heat transfer coefficient. Furthermore, the relationship between the enhancement factor and the overall heat transfer rare is extremely simple, and consequently various aspects of the overall heat transfer rate are readily deduced from the enhancement factor.

FIN ASSEMBLY FORMULATION

Assumptions and Model

Consider a heat exchanger comprised of equally spaced longitudinal rectangular fins attached to a plane wall, as depicted schematically in Fig. 1. In this study a theoretical representation of this device is developed on the basis of the following assumptions:

1. the materials from which the wall and fins are fabricated are isotropic and have constant thermal conductivities;

2. there is perfect contact between the wall and fins;

3. the convective heat transfer can be described using Newton's Law of Cooling [6];

4. the fin side heat transfer coefficient is constant over the fin surface;

5. the heat flow within the fins is one-dimensional;

6. the heat flow within the wall is one-dimensional even in the presence of the fins.

These are essentially the classical assumptions inherent in the analysis of conducting-convecting finned surfaces, e.g. see [1-7].

Several investigations [8,9,10] have indicated that the assumption (4) is not strictly valid. However, these investigations have failed to establish any generally applicable data regarding the variation of the heat transfer coefficient over the fin surface. Thus, an accurate representation of the convective heat transfer would require a simultaneous analysis of both the heat transfer and the fluid flow around the fins. Unfortunately, the mathematical complexity of such a representation is beyond the scope of the present investigation.

The justification of the assumption (5) is conventionally based on the criterion that the fin length be very much larger than the fin thickness, e.g. see [1-4]. However, several recent investigations [11,12,13] have shown that the applicability of the one-dimensional fin approximation is dependent upon the transverse Biot number Bi, and not the ratio of the fin length to the fin thickness. The inference of both these criteria is that the sum of resistances method for predicting the overall heat transfer rate is applicable provided the heat flow within the fins is effectively one-dimensional. However, since the sum of resistances method neglects the two-dimensional effects induced within the wall by the presence of the fins [14,15,16], it is apparent that its applicability can only be justified by a direct comparison with the overall heat transfer rates predicted by a completely two-dimensional analysis. Therefore, no attempt is made in this study to qualify the applicability of the one-dimensional approximation. In particular, no approximations which require the fin length to be large in comparison to the fin thickness are introduced.

In order to develop a mathematical representation it is necessary to decide which section of the fin assembly should be examined. The geometrical symmetry of the fin assembly configuration, and the thermal symmetry resulting as a consequence of the assumptions (2), (5) and (6) indicate that it will suffice to examine the heat flow within the region (A+B+C), Fig. 1. This is the smallest region which includes all the essential features of the complete fin assembly, namely, the wall, fin and interfin spacing.

It may initially appear superfluous to examine the heat flow within the region A since, by virtue of the assumption (6), the temperature distribution within this region will be identical to that within the region B. However, if attention is restricted solely to the region (B+C), then the effects of the interfin spacing will not be accounted for and therefore the model will not be representative of the fin assembly.

Mathematical Analysis

On the basis of the preceeding development, the determination of the temperature distribution within the fin assembly (and hence the heat transfer rate) requires the simultaneously solution of the energy equations ([6]),

$$\frac{d^2}{dX^2} \phi_w(X) = 0, \qquad \text{within the wall (region (A+B), Fig.1)} \qquad (1)$$

and

$$\frac{d^2}{dZ^2} \phi_f(Z) - \frac{Bi}{T^2} \phi_f(Z) = 0, \qquad \text{within the fin (region C, Fig. 1.)} \qquad (2)$$

subject to the boundary conditions,

at $X = 0$ $\qquad \frac{d}{dX} \phi_w(X) = - Bi_1 (1-\phi_w(X))$ $\qquad$ (3a)

at $X = W$ $(Z=0)$ $\qquad \phi_w(X) = \phi_f(Z)$ $\qquad$ (3b)

and $\qquad \frac{d}{dX} \phi_w(X) = \kappa T \frac{d}{dZ} \phi_f(Z) - Bi_2(1-T) \phi_w(X)$ $\qquad$ (3c)

at $Z=L$ $\qquad \frac{d}{dZ} \phi_f(Z) = -Bi_2 \phi_f(Z)$ $\qquad$ (3d)

The boundary conditions (3a) and (3d) describe the convective heat exchange at the plain side of the wall and the tip of the fin, respectively. The boundary conditions (3b) and (3c) arise as a consequence of the perfect contact assumption, and stipulate continuity of temperature and heat flux across the wall-to-fin interface.

It may appear that the boundary condition (3c) can be equivalently re-expressed in the form,

at X=W (Z=0) $\dfrac{d}{dX} \phi_w(X) = \kappa \dfrac{d}{dZ} \phi_f(Z)$, $0 \leqslant Y \leqslant T$ (4a)

and $\dfrac{d}{dX} \phi_w(X) = -Bi_2 \phi_w(X)$, $T \leqslant Y \leqslant 1$ (4b)

However, this results in a solution which, in effect, neglects the thermal interaction of the regions A, B and C, Fig. 1. The temperature distribution in region A is found to be exactly that as if the fins were not present. Furthermore, the temperature distribution in region B differs from that in region A and is unaffected by variations in the interfin spacing. Thus, it is clearly inappropriate to employ the boundary conditions (4a) and (4b).

The solution to the problem described by the equations (1), (2) and (3) is

$$\phi_w(X) = \frac{-X + W + \varepsilon \dfrac{1}{Bi_2}}{\dfrac{1}{Bi_1} + W + \varepsilon \dfrac{1}{Bi_2}}$$ (5)

and

$$\phi_f(Z) = \frac{\varepsilon \dfrac{1}{Bi_2}}{\dfrac{1}{Bi_1} + W + \varepsilon \dfrac{1}{Bi_2}} \left\{ \frac{\cosh(1-Z)\dfrac{L}{T} Bi^{\frac{1}{2}} + Bi^{\frac{1}{2}} \sinh(1-Z)\dfrac{L}{T} Bi^{\frac{1}{2}}}{\cosh \dfrac{L}{T} Bi^{\frac{1}{2}} + Bi^{\frac{1}{2}} \sinh \dfrac{L}{T} Bi^{\frac{1}{2}}} \right\}$$

(6)

where

$$\varepsilon = \frac{Bi_2}{Bi_2(1-T) + \kappa Bi^{1/2} \left\{ \frac{\sinh \frac{L}{T} Bi^{1/2} + Bi^{1/2} \cosh \frac{L}{T} Bi^{1/2}}{\cosh \frac{L}{T} Bi^{1/2} + Bi^{1/2} \sinh \frac{L}{T} Bi^{1/2}} \right\}} \qquad (7)$$

Fin Assembly Heat Transfer Rate

The heat flow rate through the fin assembly is most conveniently expressed in the form of an augmentation factor, Aug, defined as the ratio of the heat transfer rate of the finned assembly to that of the unfinned wall operating under the same conditions. In order to evaluate the augmentation factor it is first necessary to determine the heat flow rate through the unfinned wall. This requires the solution of the energy equation,

$$\frac{d^2}{dX^2} \phi_w^*(X) = 0, \qquad \text{within the wall} \qquad (8)$$

subject to the boundary conditions,

at X=0 $\qquad \frac{d}{dX}\phi_w^*(X) = -Bi_1(1-\phi_w^*(X))$ $\qquad (9a)$

at X=W $\qquad \frac{d}{dX}\phi_w^*(X) = -Bi_2 \phi_w^*(X)$ $\qquad (9b)$

where the differential equation (8) is obtained by performing an energy balance on an infinitesimal element of the wall, and the boundary conditions (9a) and (9b) describe the convective heat exchange from the surfaces of the wall.

The solution to this problem is

$$\phi_w^*(X) = \frac{-X + W + \frac{1}{Bi_2}}{\frac{1}{Bi_1} + W + \frac{1}{Bi_2}} \qquad (10)$$

and therefore,

$$A_{ug} = \frac{\frac{1}{Bi_1} + W + \frac{1}{Bi_2}}{\frac{1}{Bi_1} + W + \varepsilon\frac{1}{Bi_2}} \tag{11}$$

DISCUSSION OF FIN ASSEMBLY FORMULATION

Comparison with Sum of Resistances Method

In the sum of resistances approach the unfinned wall and the fin assembly are considered analogous to electric circuits as shown in Figs. 2a and 2b, respectively [6]. On the basis of the heat flow (i.e. "electric current") within these circuits, the augmentation factor takes the form,

$$A_{ug} = \frac{R_1 + R_w + R_2}{R_1 + R_w + \frac{1}{R_{ifs} + R_{fin}}} \tag{12}$$

where R_1 $(= 1/h_1 P)$ and R_2 $(= 1/h_2 P)$ are the convective resistances of the surfaces of the wall, R_w $(= w/k_w P)$ is the conductive resistance of the wall, R_{ifs} $(= 1/h_2(P-t))$ is the convective resistance of the interfin surface and R_{fin} $(= 1/h_2(1+t)\eta)$ is the total thermal resistance of the fin. The introduction of the appropriate expression for the fin efficiency η (e.g. as given in Kern and Kraus [7,p.170]) enables the equation (12) to be re-expressed as,

$$A_{ug} = \frac{R_1 + R_w + R_2}{R_1 + R_w + \varepsilon R_2} \tag{13}$$

$$\equiv \frac{\frac{1}{Bi_1} + W + \frac{1}{Bi_2}}{\frac{1}{Bi_1} + W + \varepsilon\frac{1}{Bi_2}}$$

where ε is defined by the equation (7). Thus, the sum of resistances method results in exactly the same solution as that given by the mathematically rigorous formulation presented in the preceding section. The principal reason for this, besides the similarity in the underlying assumptions, is the similarity in the way in which the thermal interaction between the wall and fins is represented. In the sum of resistances method, the thermal resistances of the fin (R_{fin}) and the interfin surface of the wall (R_{ifs}) both branch from the same node, Fig. 2b. Thus, by Kirchoff's Law, the total heat flow to that node is equal to the sum of the heat flow into the fin and the heat flow across the interfin surface. This is identically equivalent to the boundary condition (3c).

Enhancement-Factor

From the expression (11) (or (13)) for the augmentation factor, it is apparent that the factor ε is the only quantity which involves the fin parameters, namely, the fin length, fin thickness, fin thermal conductivity and interfin spacing. Thus, the effects of variations in the fin parameters, on the overall heat transfer rate, are indicated solely by this factor ε. This is in contrast to the conventional approach in which the fin efficiency η is employed in order to indicate the fin performance, and a separate analysis is performed in order to account for the interfin spacing. Furthermore, because of the relatively simple relationship between this factor ε and the augmentation factor, the effect of the fins on the overall heat transfer can easily be deduced from the value of ε. In particular, the overall heat transfer rate is enhanced by the addition of fins if, and only if, the fins are arranged in such a manner that the resulting value of ε is less than unity; accordingly, ε is referred to as the enhancement-factor. In addition, this enhancement-factor has the property that the smaller its value is, the greater is the corresponding augmentation. However, for any given problem, there is an upper limit on the augmentation that can be achieved by the addition of fins. This is denoted by Aug_∞ and corresponds to the ideal case in which $\varepsilon=0$ and the fin heat dissipation is infinite, namely,

$$Aug_\infty = \frac{\frac{1}{Bi_1} + W + \frac{1}{Bi_2}}{\frac{1}{Bi_1} + W} \equiv \frac{R_1 + R_w + R_2}{R_1 + R_w} \qquad (14)$$

It is not practical to design for Aug_∞ because once the fin side resistance εR_2, equation (13), is reduced to the value of the sum of the other two resistances, namely R_1 and R_w any further reductions in εR_2 will not yield any significnant increase in the heat transfer. Therefore, for sensible utilisation of fins the corresponding enhancement-factor should be such that the value of εR_2 is not less than $R_1 + R_w$, i.e. the augmentation factor should not exceed $Aug_\infty/2$.

In contrast to the useful design information provided by the enhancement factor, the fin efficiency gives no indication whatsoever regarding the effect of the fins on the overall heat transfer.

DISCUSSION OF RESULTS

From the expression (7) it can be deduced that the enhancement factor ε may be parameterised by the Biot number Bi_2, the ratio of the thermal conductivities κ, and the aspect ratios L and T (since $Bi = Bi_2 T/\kappa$). The behaviour of the enhancement factor with variations in these parameters has been extensively examined with a view to understanding the effects of changes in the fin parameters on the overall heat transfer rate. Representative results are given in in Figs. 3 and 4. In each of the graphs presented in these figures, the parameters Bi_2 and κ are assigned prescribed values and then the enhancement factor ε is plotted against L for three different values of T. The salient features of these results are discussed below:

The Effects of Variations in the Fin Length

A most prominent feature of the results is the manner in which the enhancement factor inevitably approaches some limiting value as the fin length is increased. This limiting value can in fact be determined analytically by examining the behaviour of ε for large values of L. The expression (7) for ε can be equivalently expressed as

$$\varepsilon = \frac{1}{(1-T) + \left(\frac{\kappa T}{Bi_2}\right)^{\frac{1}{2}} \left\{ \frac{\tanh \frac{L}{T} Bi^{\frac{1}{2}} + Bi^{\frac{1}{2}}}{1 + Bi^{\frac{1}{2}} \tanh \frac{L}{T} Bi^{\frac{1}{2}}} \right\}} \qquad (15)$$

Therefore, for large values of L,

$$\varepsilon \sim \frac{1}{(1-T) + \left(\frac{\kappa T}{Bi_2}\right)^{\frac{1}{2}}} \qquad (16)$$

In all cases the results presented in Figs. 3 and 4 agree with the limiting value predicted by the equation (16).

This tendency of the enhancement-factor to approach some limiting value indicates that the overall heat transfer cannot be indefinitely increased by increasing the fin length. It is not immediately apparent why this should be the case. However, an explanation can be formulated from a physical viewpoint. The addition of fins to a plane wall results in an increase in the total heat transfer surface area, but at the same time introduces an additional conductive resistance. Initially the gain in surface area far outweighs the extra conductive resistance. However, eventually a state is reached such that further increases in the fin length are negated by the respective increases in the conductive resistance.

The Effects of Variation in the Fin Thickness and Fin Pitch

Another prominent aspect of the results presented in Figs. 3 and 4 is the feature that as the Biot number Bi_2 is reduced, the variation of the enhancement-factor with the fin thickness is reduced. In practical terms, reductions in Bi_2 correspond to reductions in the fin pitch P. A reduction of the fin pitch proportionately reduces the possible variations in the fin thickness and therefore the effect on the enhancement-factor is less marked.

A particularly significant deficiency of the one-dimensional analysis is highlighted in the Figs. 3 and 4. The case $T = 1.00$ corresponds to the situation in which the interfin spacing is zero, and therefore, the fin material effectively increases the wall thickness. This should result in a reduction of the overall heat flow rate and consequently the corresponding value of ε should be greater than unity. However, the results show that the corresponding values of ε are not only less than unity, but are also less than those for other values of the parameter T. Thus, the one-dimensional analysis is grossly inappropriate for situations in which the interfin spacing is much less than the fin thickness.

The Effects of Variation in the Fin Thermal Conductivity

The effects of variations in the fin thermal conductivity on the overall heat transfer rate can be deduced by comparing the results for different values of κ. For prescribed fin dimensions, interfin spacing and surface heat transfer coefficient, i.e. for fixed values of L,T,P and Bi_2, an increase in the fin thermal conductivity, i.e. a higher value of κ, results in a reduction in the value of the enhancement-factor, e.g. see Figs. 3 and 4. This reduction can be quite substantial for larges values of Bi_2. Thus, as would be expected, increasing the fin thermal conductivity facilitates an increase in the overall heat flow rate.

CONCLUSIONS

A mathematically rigorous formulation has been developed for analysing the heat flow within finned surfaces. The solutions predicted by this formulation are identical to those obtained employing the conventional (mathematically informal) sum of resistances method. However, this new formulation has several advantageous design properties in comparison with the conventional method. In particular, a new factor ε is introduced which accounts for the effects of fin dimensions, interfin spacing and fin thermal conductivity, and thereby avoids the need to perform separate design calculations for the various components of the fin assembly, as is necessary with the conventional method.

It must be emphasised that the formulation presented in this paper is equally applicable to other fin profiles and to the annular geometry. In fact, the expressions for the enhancement-factor for (a) longitudinal rectangular fins attached to a plane wall, (b) longitudinal triangular fins attached to a plane wall, (c) annular rectangular fins attached to a cylindrical tube, and (d) annular triangular fins attached to a cylindrical tube, are tabulated in the Appendix in a form convenient for design purposes, and a complete set of enhancement-factor design charts for these four configurations may be obtained from the authors.

NOMENCLATURE

Aug   augmentation factor

Aug_∞ ideal maximum augmentation factor

Bi $= h_2 t/k_f$, Biot number

Bi_1 $= h_1 P/k_w$, Biot number

Bi_2 $= h_2 P/k_w$, Biot number

h_1, h_2 heat transfer coefficients, W/m²K

k_f, k_w thermal conductivities, W/mK

l   fin length, m

L $= l/P$, aspect ratio

P   half fin pitch, m

t   half fin thickness, m

T $= t/P$, aspect ratio

w   wall thickness, m

W $= w/P$, aspect ratio

x   longitudinal displacement in wall, m

X $= x/P$

y   transverse displacement, m

Y $= y/P$

z   longitudinal displacement in fin, m

Z $= z/P$

ε enhancement-factor

η fin efficiency

κ $= k_f/k_w$

θ temperature distribution, K

θ_1, θ_2 fluid temperatures, K

ϕ $= (\theta-\theta_2)/(\theta_1-\theta_2)$, dimensionless temperature distribution

Subscripts

1   plain side

2   fin side

f   fin

w   wall

Superscript

*   unfinned wall

REFERENCES

1. D.R. Harper and W.B. Brown, "Mathematical equations for heat conduction in the fins of air-cooled engines", National Advisory Committee for Aeronautics, Report 158, 1922.

2. K.A. Gardner, "Efficiency of extended surface", Transactions of the ASME, Vol. 67, pp. 621-631, 1945.

3. S. Guceri and C.J. Maday, "A least weight circular cooling fin", Journal of Engineering for Industry, Vol. 97, pp. 1190-1193, 1975.

4. I. Mikk, "Convective fin of minimum mass", International Journal of Heat and Mass Transfer, Vol. 23, pp. 707-711, 1981.

5. P.J. Heggs, D.B. Ingham and M. Manzoor, "The effects of non-uniform heat transfer from an annular fin of triangular profile", Journal of Heat Transfer, Vol.103, pp. 184-185, 1981.

6. F. Kreith, Principles of heat transfer , Harper and Row, New York, 1976.

7. D.Q. Kern and A.D. Kraus, Extended surface heat transfer , McGraw-Hill, New York, 1972.

8. P.W. Wong, "Mass and heat transfer from circular finned cylinders", Journal of the Institution of Heating and Ventilating Engineers, Vol. 23, pp. 1-23, 1963.

9. J.W. Stachiewicz, "Effect of variation of local film coefficient on fin performance", Journal of Heat Transfer, Vol. 91, pp. 21-26, 1969.

10. P.J. Heggs and P.R. Stones, "Improved design methods for finned tube heat exchangers", Transactions of the Institution of Chemical Engineers, Vol. 58, pp. 147-154, 1980.

11. R.K. Irey, "Errors in the one-dimensional fin solution", Journal of Heat Transfer, Vol. 90, pp. 175-176, 1968.

12. M. Levitsky, "The criterion for the validity of the fin approximation", International Journal of Heat and Mass Transfer, Vol. 15, pp. 1960-1963, 1972.

13. W. Lau and C.W. Tan, "Errors in one-dimensional heat transfer analysis in straight and annular fins", Journal of Heat Transfer, Vol. 95, pp. 549-551, 1973.

14. E.M. Sparrow and L. Lee, "Effects of fin-base temperature depression in a multifin array", Journal of Heat Transfer, Vol. 97, pp. 463-465, 1975.

15. N.V. Suryanarayana, "Two-dimensional effects on heat transfer from an array of straight fins", Journal of Heat Transfer, Vol. 99, pp. 129-132, 1977.

16. P.J. Heggs and P.R. Stones, "The effects of dimensions on the heat flowrate through extended surfaces", Journal of Heat Transfer, Vol. 102, pp. 180-182, 1980.

APPENDIX: Expressions for the enhancement-factor ε

Longitudinal fins attached to a plane wall

(a) rectangular profile	$Bi_2 / \{ Bi_2(1-T) + \kappa\, Bi^{\frac{1}{2}} \frac{(\tanh\xi + Bi^{\frac{1}{2}})}{(1+Bi^{\frac{1}{2}}\tanh\xi)} \}$	where $\xi = \frac{L}{T} Bi^{\frac{1}{2}}$
(b) triangular profile	$Bi_2 / \{ Bi_2(1-T) + \kappa\, \frac{Bi}{\cos\alpha}\, \frac{I_1(\xi)}{I_0(\xi)} \}$	where $\xi = 2 \frac{L}{T} \frac{Bi}{\cos\alpha}^{\frac{1}{2}}$

Annular fins attached to a cylindrical tube

(c) rectangular profile *	$Bi_2 / \{ Bi_2(1-T) + \kappa\, Bi^{\frac{1}{2}} \frac{(K_1(\xi)I(\xi) - I_1(\xi)K(\eta))}{(K_0(\xi)I(\eta) + I_0(\xi)K(\xi))} \}$
	where $\xi = \frac{R_b}{T} Bi^{\frac{1}{2}}$, $\eta = \frac{R_t}{T} Bi^{\frac{1}{2}}$, $I(\eta) = I_1(\eta) + Bi^{\frac{1}{2}} I_0(\eta)$ and $K(\eta) = K_1(\eta) - Bi^{\frac{1}{2}} K_0(\eta)$
(d) triangular profile *	$Bi_2 / \{ Bi_2(1-T) + \kappa \frac{T}{L} (\sum_{n=1}^{\infty} n\, a_n / \sum_{n=0}^{\infty} a_n) \}$
	where $a_0 = 1$, $a_1 = \xi$, $a_2 = ((n(n+1) + \xi n)a_{n-1} - \xi a_n)/n^2 \eta$, $n \geq 2$
	with $\xi = \left(\frac{L}{T}\right)^2 \frac{Bi}{\cos\alpha}$ and $\eta = \frac{Rt}{L}$

* $R_b = r_b/P$, $R_t = r_t/P$; where r_b is the radius of the fin base and r_t is the radius of the fin tip

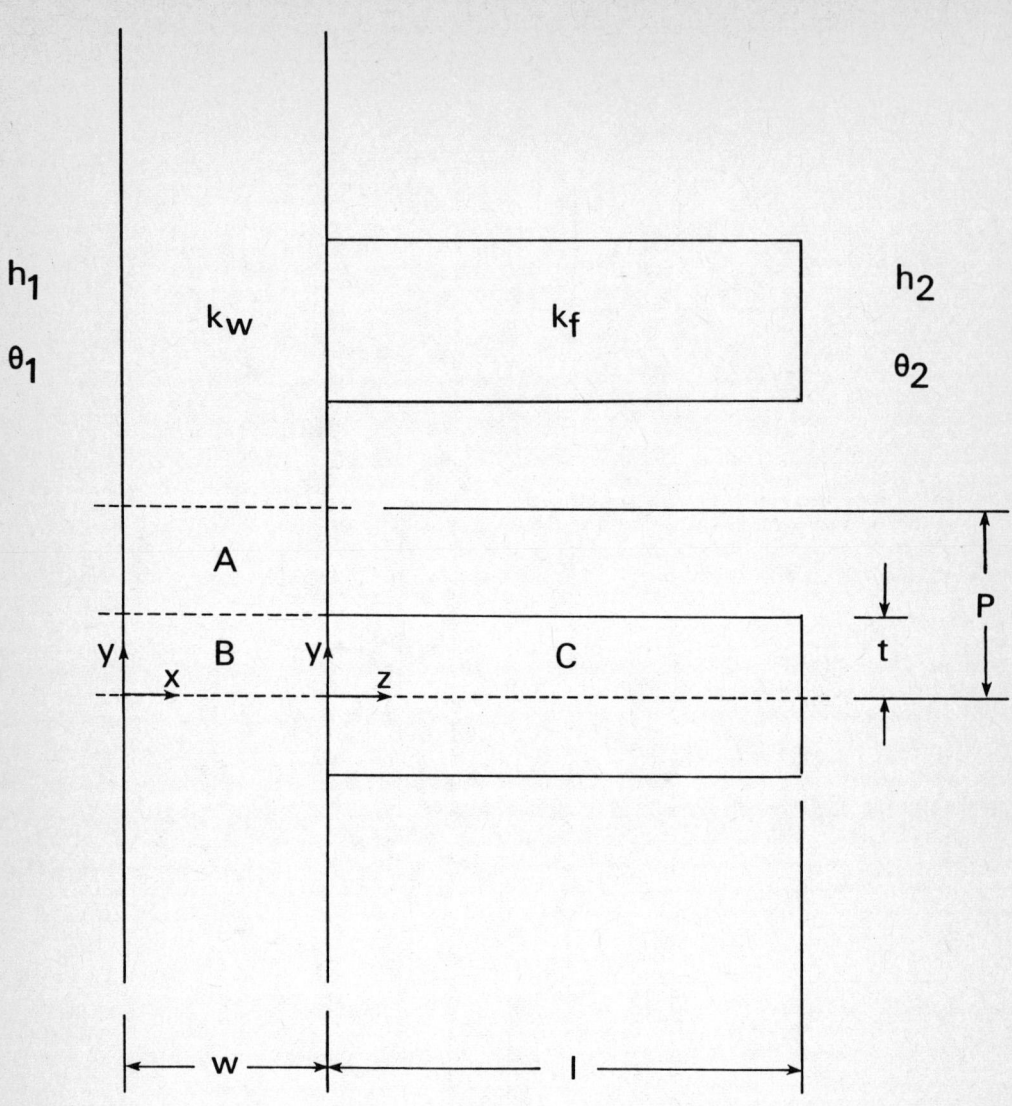

Fig.1 Schematic representation of a fin assembly

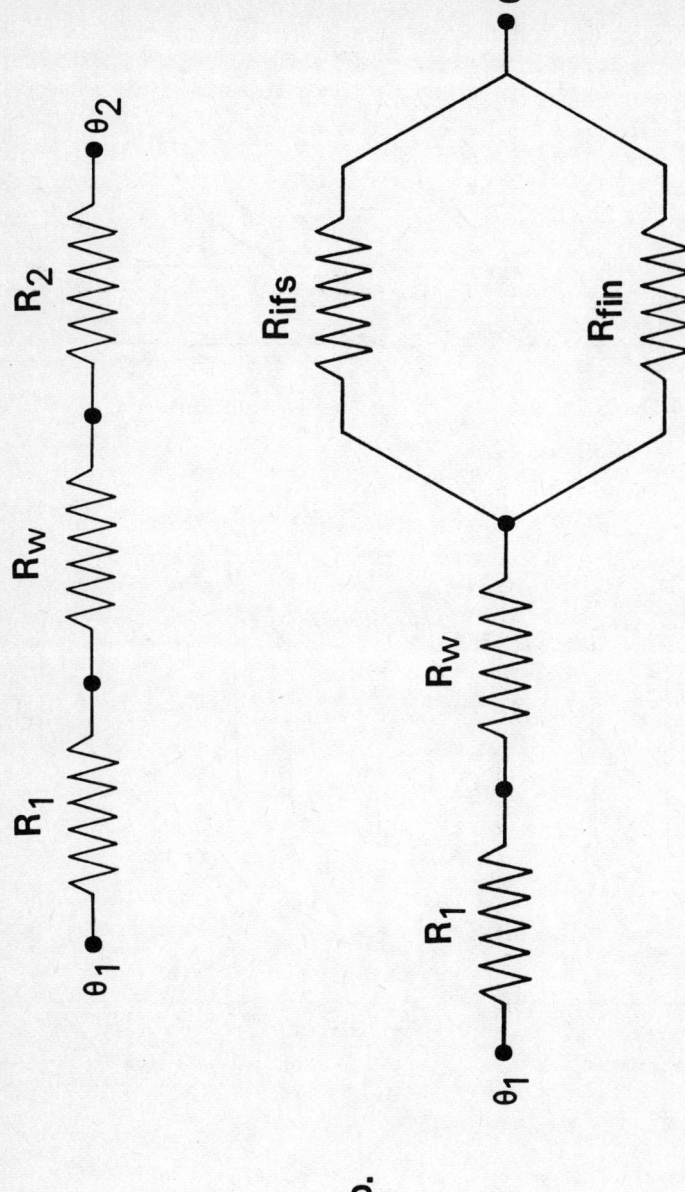

Fig.2 Thermal curcuit representations for the heat flow within (a) the unfinnned wall, and (b) the fin assembly

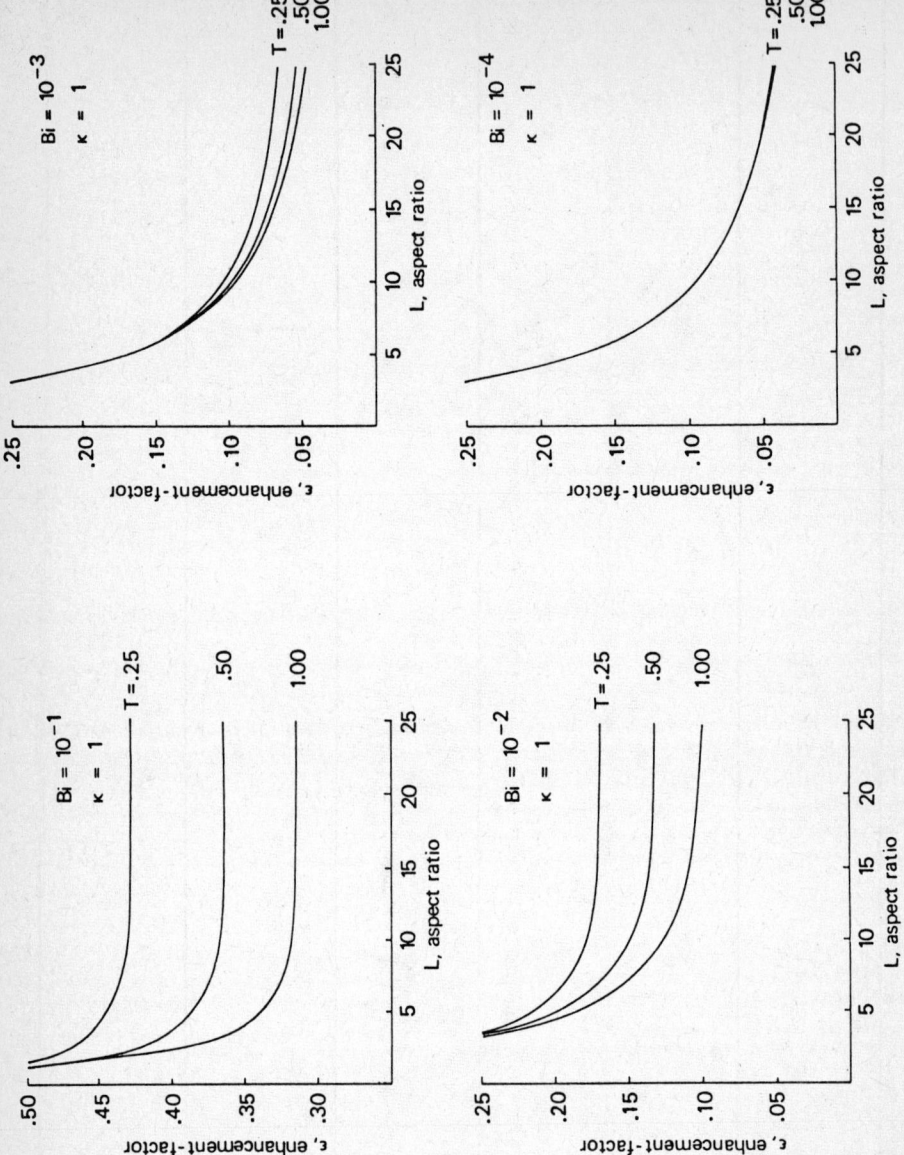

Fig.3 The effects of variations in Bi_2, L and T on the enhancement-factor ε for the case $\kappa = 1$

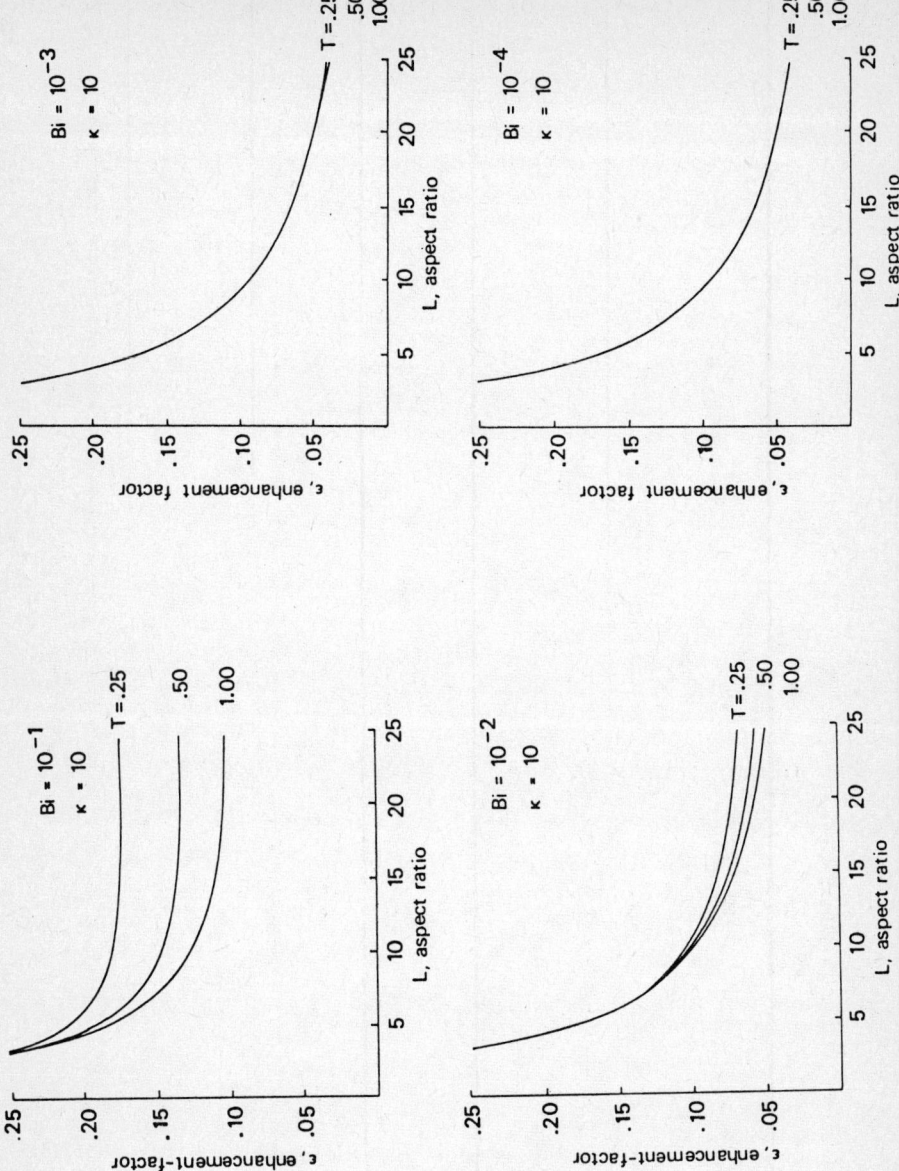

Fig. 4 The effects of variations in Bi_2, L and T on the enhancement-factor ε for the case $\kappa = 10$

CHAPTER 3

THE TWO-DIMENSIONAL ANALYSIS OF FIN ASSEMBLY HEAT TRANSFER

3.1 THE NUMERICAL SOLUTION OF PLANE POTENTIAL PROBLEMS BY IMPROVED BOUNDARY INTEGRAL EQUATION METHODS

ABSTRACT

In this study a modified boundary integral equation method which enables the accurate solution of Laplacian mixed boundary-value problems is presented. This method is designed specifically for the treatment of problems in which boundary singularities occur on the interface between two regions which have different physical properties. The advanced solution capabilities of this method are illustrated by the application to two physical problems. In addition, previously undetermined analytical expressions for the kernel integrals arising in the piecewise-linear and piecewise-quadratic BIE implementations are also presented. In comparison with the previously employed quadrature formulae these analytical expressions afford an appreciable reduction in the computational time.

INTRODUCTION

Elliptic boundary-value problems arising from the examination of physical situations, such as those encountered in engineering and mathematical physics, are, in general, intractable by analytical treatment. The solution to such problems can usually be computed by numerical techniques such as the finite-difference [1], finite-element [2] and boundary integral equation [3] methods. However, the standard forms of these numerical techniques tend to yield inaccurate solutions for problems involving boundary singularities. Consequently, the possibility of modifying the standard techniques in order to give special treatment to the singular points and thereby to obtain solutions which converge more rapidly has received considerable attention, e.g. [4-10].

Symm [6] has devised a modified BIE formulation which can successfully treat boundary singularities in two-dimensional Laplacian problems. The results obtained employing this modified BIE method offer considerable improvement over those given by finite-difference and finite-element methods modified by either mesh refinement in the neighbourhood of the singularity or inclusion of terms having the analytical form of the singularity. In this study the modified BIE method is extended in order to enable the accurate solution of problems in which the boundary singularity occurs on the interface between two regions which have different physical properties, e.g. different dielectric permittivities in electromagnetics [4] and different thermal conductivities in heat diffusion problems [11]. In order to illustrate the solution capabilities of this method two problems which involve L-shaped domains with mixed boundary conditions are examined and solutions are contrasted with those obtained by employing standard piecewise-constant, piecewise-linear and piecewise-quadratic BIE implementations. Furthermore, previously undetermined analytical expressions for the kernal integrals associated with the piecewise-linear and piecewise-quadratic BIE formulations are presented. The use of these analytical expressions, instead of the previously employed quadrature formulae [13], not only reduces the programming complexity but also results in a substantial reduction in the computational time.

THE STANDARD BIE METHODS

Formulation

There is an extensive range of published literature giving detailed descriptions of the various BIE formulations for obtaining solutions to plane potential boundary-value problems, e.g. [3,6,7,12,13]. Consequently, only those features necessary to facilitate a concise explanation of the proposed modification are presented in this study.

The fundamental basis of the BIE method is Green's Integral Formula [3] which, for any sufficiently smooth function ϕ which satisfies Laplace's equation in a plane domain Ω having a piecewise-smooth boundary $\partial\Omega$, may be expressed as,

$$\int_{\partial\Omega} \{\phi(q) \log'|p-q| - \phi'(q) \log |p-q|\} dq = \eta(p) \phi(p) \tag{1}$$

where,

i $p \in \Omega + \partial\Omega$, $q \in \partial\Omega$

ii dq denotes the differential increment of $\partial\Omega$ at q

iii   the prime (') denotes the derivative in the direction of the outward normal to $\partial\Omega$ at q

iv   if $p \in \Omega$ then $\eta = 2\pi$, and if $p \in \partial\Omega$ then η is the angle included between the tangents to $\partial\Omega$ on either side of p.

If either ϕ, ϕ' or a linear combination of ϕ and ϕ' is prescribed at each point of $\partial\Omega$ then the solution of the integral equation,

$$\int \{\phi(q) \log'|\bar{q}-q| - \phi'(q) \log |\bar{q}-q|\} dq - \eta(\bar{q}) \phi(\bar{q}) = 0, \quad q,\bar{q} \in \partial\Omega \tag{2}$$

determines ϕ and ϕ' at each point of $\partial\Omega$. The potential ϕ at any point

$p \in (\Omega+\partial\Omega)$ can then be computed employing Green's Integral Formula, equation (1).

Thus, the application of Green's Boundary Formula, equation (2), enables a well-posed two-dimensional Laplacian boundary-value problem to be reformulated as an integral equation in which the unknowns are the boundary values of the potential ϕ and its normal derivative ϕ' complementary to those prescribed by the boundary conditions. In practice, the integral equation (2) can rarely be solved analytically. Consequently, various numerical techniques have been proposed in order to enable the determination of a solution.

In the classical BIE (CBIE) method [13] the boundary $\partial\Omega$ is first subdivided into N smooth intervals $\partial\Omega_j$, j=1,...,N, on which ϕ and ϕ' are approximated by piecewise-constant functions ϕ_j and ϕ'_j. Then, the corresponding discretized form of Green's Integral Formula,

$$\sum_{j=1}^{N} \{\phi_j \int_{\partial\Omega_j} \log'|p-q|dq - \phi'_j \int_{\partial\Omega_j} \log|p-q|dq\} = \eta(p)\phi(p),$$
$$p \in \Omega + \partial\Omega, \; q \in \partial\Omega \qquad (3)$$

is collocated at the midpoint q_j of each interval.

This generates a system of linear algebraic equations in the unknown ϕ_j and ϕ'_j. Solution of this system of linear algebraic equations determines the values of both ϕ_j and ϕ'_j on each interval. The potential ϕ at any interior point may then be computed by a relatively simple quadrature, equation (3), if required.

The linear BIE (LBIE) method [12] affords a slightly more sophisticated approximation of Green's Integral Formula than the classical BIE method. On each interval $\partial\Omega_j$, j=1,...,N, ϕ and ϕ' are approximated by piecewise-linear functions

$$\phi = (1 - \xi) \phi(q_j) + \xi\phi(q_{j+1})$$

$$\phi' = (1 - \xi) \phi'(q_j) + \xi\phi'(q_{j+1})$$

where q_j and q_{j+1} are the endpoints of $\partial\Omega_j$, and ξ is a linear function which increases from zero at q_j to unity at q_{j+1}. With these approximations Green's Integral Formula becomes,

$$\sum_{j=1}^{N} \{\phi_j \int_{\partial\Omega_j} (1-\xi) \log'|p-q|dq + \phi_{j+1} \int_{\partial\Omega_j} \xi \log'|p-q|dq\}$$

$$- \sum_{j=1}^{N} \{\phi'_j \int_{\partial\Omega_j} (1-\xi) \log|p-q|dq + \phi'_{j+1} \int_{\partial\Omega_j} \xi \log|p-q|dq\}$$

$$= \eta(p) \phi(p), \qquad p \in \Omega + \partial\Omega, q \in \partial\Omega \qquad (4)$$

where ϕ_j and ϕ'_j denote $\phi(q_j)$ and $\phi'(q_j)$, respectively. A system of linear algebraic equations in the unknown ϕ_j and ϕ'_j is then generated by enforcing equation (4) at each of the points q_j.

More accurate approximation of the solution to the boundary integral equation can be obtained using the quadratic BIE (QBIE) method [13]. In this approach, on each interval $\partial\Omega_j$, $j=1,\ldots,N, \phi$ and ϕ' are approximated by piecewise-quadratic functions,

$$\phi = M_1(\xi) \phi(q_{2j-1}) + M_2(\xi) \phi(q_{2j}) + M_3(\xi) \phi(q_{2j+1})$$

$$\phi' = M_1(\xi) \phi'(q_{2j-1}) + M_2(\xi) \phi'(q_{2j}) + M_3(\xi) \phi'(q_{2j+1})$$

where q_{2j-1} and q_{2j+1} are the endpoints of $\partial\Omega_j$, q_{2j} is the midpoint of $\partial\Omega_j$, ξ is a linear function which increases from zero at q_{2j-1} to unity at q_{2j+1}, and

$$M_1(\xi) = 1 - 3\xi + 2\xi^2$$
$$M_2(\xi) = 4\xi - 4\xi^2$$
$$M_3(\xi) = -\xi - 2\xi^2$$

On the basis of these approximations Green's Integral Formula becomes,

$$\sum_{j=1}^{N} \{\phi_{2j-1} \int_{\partial\Omega_j} M_1(\xi) \log'|p-q| dq + \phi_{2j} \int_{\partial\Omega_j} M_2(\xi) \log'|p-q| dq$$

$$+ \phi_{2j+1} \int_{\partial\Omega_j} M_3(\xi) \log'|p-q| dq\}$$

$$- \sum_{j=1}^{N} \{\phi'_{2j-1} \int_{\partial\Omega_j} M_1(\xi) \log|p-q| dq + \phi'_{2j} \int_{\partial\Omega_j} M_2(\xi) \log|p-q| dq$$

$$+ \phi'_{2j+1} \int_{\partial\Omega_j} M_3(\xi) \log|p-q| dq\}$$

$$= \eta(p) \phi(p), \quad p \in \Omega + \partial\Omega, \ q \in \partial\Omega. \tag{5}$$

A system of linear algebraic equations in the unknown ϕ_j and ϕ'_j is then generated by applying formula (5) to each of the points q_j, $j=1,\ldots,2N$. Thus, for an N interval discretization, the QBIE method requires the solution of 2N equations in 2N unknowns, whereas the CBIE and LBIE methods only require the solution of N equations in N unknowns.

Kernel Functions

If the interval $\partial\Omega_j$ is a straight line segment, then the integrals occurring in the formulae (3), (4) and (5) can be evaluated exactly using,

$$\int_{\partial\Omega_j} \log'|p-q| dq = I_1 \tag{6}$$

$$\int_{\partial\Omega_j} \log|p-q| dq = J_1 \tag{7}$$

$$\int_{\partial\Omega_j} \xi \log'|p-q| dq = \frac{1}{h}(a \cos \beta \ I_1 + I_2) \tag{8}$$

$$\int_{\partial\Omega_j} \xi \log|p-q| dq = \frac{1}{h} (a \cos \beta \, J_1 + J_2) \qquad (9)$$

$$\int_{\partial\Omega_j} \xi^2 \log'|p-q| dq = \frac{1}{h^2} ((h-2a\cos\beta)^2 J_1 - 4(h-2a\cos\beta) J_2 + 4J_3) \qquad (10)$$

$$\int_{\partial\Omega_j} \xi^2 \log|p-q| dq = \frac{1}{h^2} ((h-2a\cos\beta)^2 J_1 - 4(h-2a\cos\beta) J_2 + 4J_3) \qquad (11)$$

where,

$$I_1 = \psi \qquad (12)$$

$$I_2 = a \sin \beta (\log b - \log a) \qquad (13)$$

$$I_3 = a \sin \beta (h - a \psi \sin \beta) \qquad (14)$$

$$J_1 = a \cos \beta (\log a - \log b) + h (\log b - 1) + a \psi \sin\beta \qquad (15)$$

$$J_2 = \frac{1}{2} (b^2 \log b - a^2 \log a) - \frac{1}{4} (b^2 - a^2) \qquad (16)$$

$$J_3 = \frac{1}{3} \{ (h - a \cos \beta)^3 (\log b - \frac{1}{3}) + (a \cos \beta)^3 (\log a - \frac{1}{3})$$
$$+ (a \sin \beta)^2 (h - a \psi \sin \beta) \} \qquad (17)$$

and if q_{aj} and q_{bj} denote the endpoints of $\partial\Omega_j$, Fig. 1, then a,b and h are the lengths of the lines joining p to q_{aj}, p to q_{bj} and q_{aj} to q_{bj}, respectively, and β and ψ are the angles $q_{bj} q_{aj}$ p and q_{aj} p q_{bj}, respectively.

The analytical expressions (6) and (7) for the integrals associated with the CBIE method, equation (3), were originally presented by Symm [6]. However, the expressions (8) - (11) have not previously been determined; the integrals associated with the LBIE and QBIE methods have previously been computed numerically.

Evaluation of the integrals occurring in LBIE and QBIE formulations by the analytical expression (6) – (11) requires only a fraction of the computational time taken by an accurate numerical technique, and since, for an N interval discretization, each of the integrals has to be evaluated N times for every point to which Green's Integral Formula is applied, it is apparent that the use of these analytical expressions will facilitate appreciable reductions in the computational times required by the LBIE and QBIE methods.

Application

In order to demonstrate the problems caused by the presence of boundary singularities, the CBIE, LBIE and QBIE methods are applied to two physical problems involving L-shaped domains.

Problem 1

This problem arises in the study of plane potential flow through a porous medium between impervious pins [6], and involves an L-shaped domain, Fig. 2, with a singularity at the re-entrant corner O. The determination of the potential ϕ requires the solution of,

$$\nabla^2 \phi = 0 \quad \text{in region (A+B), Fig. 2} \tag{18}$$

subject to the boundary conditions,

$$\text{on OA} \quad \phi' = 0 \tag{19a}$$

$$\text{on AB} \quad \phi = 0 \tag{19b}$$

$$\text{on BD} \quad \phi' = 0 \tag{19c}$$

$$\text{on DE} \quad \phi = 1 \tag{19d}$$

$$\text{on EF} \quad \phi' = 0 \tag{19e}$$

$$\text{on FO} \quad \phi' = 0 \tag{19f}$$

Although the regions A and B are occupied by the same medium, we shall use this problem as a test case for the proposed modification by treating it in the form,

$$\nabla^2 \phi_A = 0 \quad \text{in region A, Fig. 2,} \tag{20}$$

$$\nabla^2 \phi_B = 0 \quad \text{in region B, Fig. 2,} \tag{21}$$

subject to the conditions,

on OA $\quad \phi_A' = 0$ \hfill (22a)

on AB $\quad \phi_A = 0$ \hfill (22b)

on BC $\quad \phi_A' = 0$ \hfill (22c)

on CO $\quad \phi_A = \phi_B$ \hfill (22d)

and $\quad -k_B \phi_B' = k_A \phi_A'$ \hfill (22e)

on CD $\quad \phi_B' = 0$ \hfill (22f)

on DE $\quad \phi_B = 1$ \hfill (22g)

on EF $\quad \phi_B' = 0$ \hfill (22h)

on FO $\quad \phi_B' = 0$ \hfill (22i)

Applying Green's Boundary Formula, equation (2), to this problem gives rise to a pair of coupled integral equations involving contour integrals around $\partial\Omega_A$ (= OABCO) and $\partial\Omega_B$ (= OCDEFO); the coupling arises through the interface boundary conditions (22d) and (22e). Solution of these integral equations, by the numerical techniques described above, determines the boundary distributions of both ϕ and ϕ'. Then in order to compute the potential ϕ at any interior point, it is only necessary to apply Green's Integral Formula to the boundary of the region in which that point lies. In particular, the potential at points on the common interface OC can be evaluated by applying Green's Integral Formula to either $\partial\Omega_A$ or $\partial\Omega_B$.

Solutions to the problem described by equations (20), (21) and (22) have been obtained by the application of the CBIE, LBIE and QBIE methods, employing 50, 100 and 200 equal length boundary intervals, for the case $OA = AB = EF = FO = 1$ and $k_A = k_B = 1$. The results presented in Tables 1a, 1b and 1c show the potential at the lattice points of a unit mesh. These results show that the LBIE and QBIE methods not only afford a better match at the common interface OC, but are also more accurate on the boundaries AB and DE, than the CBIE method.

This problem has a boundary singularity at the re-entrant corner O, Fig. 2, and the slow convergence caused by the presence of this singularity is clearly evident in the solutions displayed in Tables 1a, 1b and 1c.

Problem 2

This problem arises from the examination of the heat flow through finned surfaces [11] and involves an L-shaped composite of two rectangular domains having different thermal conductivities, Fig. 2. The temperature distribution within the domain (A+B), Fig. 2, is determined by simultaneously solving,

$$\nabla^2 \phi_A = 0 \text{ in region A, Fig. 2,} \qquad (23)$$

and $\nabla^2 \phi_B = 0$ in region B, Fig. 2, (24)

subject to the boundary conditions,

on OA $- k_A \phi_A' = h_2 \phi_A$ (25a)

on AB $- k_A \phi_A' = h_2 \phi_A$ (25b)

on BC $\phi_A' = 0$ (25c)

on CO $\phi_A = \phi_B$ (25d)

and $-k_B \phi_B' = k_A \phi_A'$ (25e)

on CD $\phi_B' = 0$ (25f)

on DE $k_B \phi_B' = h_1 (1-\phi_B)$ (25g)

on EF $\phi_B' = 0$ (25h)

on FO $-k_B \phi_B' = h_2 \phi_B$ (25i)

where k_A and k_B are the thermal conductivities of the regions A and B, respectively, and h_1 and h_2 are the heat transfer coefficients at the surfaces DE and FOAB, respectively.

One of the quantities of physical importance in this problem is the rate of heat transfer Q, which is given by [11],

$$Q = h_1 \int_{DE} (1 - \phi_B(q)) dq \qquad (26)$$

$$\equiv h_2 \{ \int_{FO} \phi_B(q) dq + \int_{OAB} \phi_A(q) dq \} \qquad (27)$$

It is apparent from the expressions (26) and (27) that evaluation of Q only requires the boundary distribution of ϕ; this is precisely the information obtained when the boundary integral equation representing the problem described by equations (23), (24) and (25) is solved.

Results have been obtained by the application of the CBIE, LBIE and QBIE methods, employing 50, 100 and 200 equal length boundary intervals, for the case OA = AB = EF = FO = 1, h_1 = 1000, h_2 = 10, k_A = 250 and k_B = 10. This represents a heat exchanger comprised of copper in region A and steel in region B, with forced convection of water along DE and free convection of air around FOAB. In Tables 2a, 2b and 2c, Q_1 and Q_2 represent the heat transfer rates corresponding to the expressions (26) and (27), respectively; as the CBIE, LBIE and QBIE methods are based on assumed boundary variations of ϕ and ϕ', they need not give the same values for Q_1 and Q_2, although obviously for the solutions to be satisfactory, Q_1 and Q_2 should agree to within an accetapble tolerance.

This problem also has a boundary singularity at the re-entrant corner O, Fig. 2, and the results displayed in Tables 2a, 2b and 2c clearly illustrate the slow convergence caused by the presence of this singularity.

In the next section a modified BIE (MBIE) method is described which gives special treatment to the singular points and thereby enables, in general, considerably more accurate solutions than those given by the standard BIE methods.

THE MODIFIED BIE METHOD

Formulation and Application

Symm [6] showed that by including terms having the analytical form of the singularity into the CBIE method, the problems caused by the presence of the singularity can be overcome. However, the method devised by Symm [6] is not directly applicable to problems in which the boundary singularity occurs on the interface between two regions which have different physical properties because the analytical solution in the neighbourhood of the singularity is represented by different expressions in the two regions. The modifications necessary to overcome this difficulty are now described for the case of problem 1. The analysis for problem 2 is very similar and therefore is not shown here.

In order to apply the MBIE method it is first necessary to determine the anlytical form of the solution in the neighbourhood of the singular point O. Employing the polar coordinates (r,ξ) in region A, and (r,η) in region B, Fig. 3, the general solutions of equations (20) and (21) can be expressed as,

$$\phi_A(r,\xi) = \sum_{n=0}^{\infty} r^{\lambda_n} (a_n \cos \lambda_n \xi + b_n \sin \lambda_n \xi) \tag{28}$$

$$\phi_B(r,\eta) = \sum_{n=0}^{\infty} r^{\mu_n} (c_n \cos \mu_n \eta + d_n \sin \mu_n \eta) \tag{29}$$

where the eigenvalues λ_n and μ_n, and the coefficients a_n, b_n, c_n and d_n are undetermined constants dependent upon the boundary conditions.

In the neighbourhood of the singularity the solutions (28) and (29) are subject to the boundary conditions,

on $\xi = 0$ $\quad\quad\quad \phi_A' = 0$ $\quad\quad$ (30a)

on $\xi = \frac{\pi}{2}$ $(\eta = \pi)$ $\quad \phi_A = \phi_B$ $\quad\quad$ (30b)

on $\xi = \frac{\pi}{2}$ $(\eta = \pi)$ $\quad -k_B \phi_B' = k_A \phi_A'$ $\quad\quad$ (30c)

on $\eta = 0$ $\quad\quad\quad \phi_B' = 0$ $\quad\quad$ (30d)

where $\xi=0$, $\eta=0$ and $\xi = \frac{\pi}{2}$ $(\eta=\pi)$ specify the boundaries OA, OF and OC respectively.

Enforcing conditions (30a) and (30d), and then matching at the common interface, using conditions (30b) and (30c), gives

$$\phi_A(r,\xi) = \alpha + \beta r^{\lambda_1} \cos \lambda_1 \xi + \gamma r^{\lambda_2} \cos \lambda_2 \xi + \delta r^{\lambda_3} \cos \lambda_3 \xi + \ldots \tag{31}$$

and

$$\phi_B(r,\eta) = \alpha^* + \beta^* r^{\lambda_1} \cos \lambda_1 \eta + \gamma^* r^{\lambda_2} \cos \lambda_2 \eta + \delta^* r^{\lambda_3} \cos \lambda_3 \eta + \ldots \tag{32}$$

where α, β, γ and δ are unknown constants, and

$$\alpha^* = \alpha$$

$$\beta^* = \beta \cos \lambda_1 \pi / \cos \lambda_1 \frac{\pi}{2}$$

$$\gamma^* = \gamma \cos \lambda_2 \pi / \cos \lambda_2 \frac{\pi}{2}$$

$$\delta^* = -\delta$$

$$\lambda_n = 2\varepsilon,\ 2(1-\varepsilon),\ 2,\ 2(1+\varepsilon),\ 2(2-\varepsilon),\ldots,\qquad n=1,2,\ldots$$

and

$$\varepsilon = \frac{1}{\pi} \cos^{-1} \left[\frac{k_A}{2(k_A+k_B)} \right]^{\frac{1}{2}}$$

Inclusion of terms of the singular solutions (31) and (32) into the CBIE method is performed by analogy with the method presented by Symm [6]. Functions ψ_A and ψ_B are defined such that,

$$\phi_A(p) = \psi_A(p) + f_A(p), \qquad p \in A + \partial A \qquad (33)$$

and

$$\phi_B(p) = \psi_B(p) + f_B(p), \qquad p \in B + \partial B \qquad (34)$$

where

$$f_A(p) = \alpha + \beta r^{\lambda_1} \cos \lambda_1 \xi + \gamma r^{\lambda_2} \cos \lambda_2 \xi + \delta r^{\lambda_3} \cos \lambda_3 \xi, \qquad (35)$$

and

$$f_B(p) = \alpha^* + \beta^* r^{\lambda_1} \cos \lambda_1 \eta + \gamma^* r^{\lambda_2} \cos \lambda_2 \eta + \delta^* r^{\lambda_3} \cos \lambda_3 \eta,$$

$$(36)$$

Thus, the functions ψ_A and ψ_B are harmonic in the regions A and B, respectively, and satisfy the boundary conditions, on OA $\quad \psi_A' = -f_A'$ (37a)

on AB $\quad \psi_A = -f_A$ (37b)

on BC $\quad \psi_A' = -f_A'$ (37c)

on CO $\quad \psi_A = -\psi_B$ (37d)

and $\quad -k_B \psi_B' = k_A \psi_A'$ (37e)

on CD $\quad \psi_B' = -f_B'$ (37f)

on DE $\quad \psi_B = -1 - f_B$ (37g)

on EF $\quad \psi_B' = -f_B'$ (37h)

on FO $\quad \psi_B' = -f_B'$ (37i)

Applying the CBIE method to the functions ψ_A and ψ_B and enforcing the boundary conditions (37) generates a system of N linear algebraic equations in N+4 unknowns, including the constants α, β, γ, and δ. In order to reduce the number of unknowns to N it is necessary to assume that $\psi_A=0$ on the intervals 1 and 2, Fig. 4, and $\psi_B=0$ on the intervals N and N-1, i.e. it is assumed that in the vicinity of the singular point O, the potentials ϕ_A and ϕ_B can be approximated by the expressions (35) and (36) for f_A and f_B. Solving this system of equations determines the boundary distributions of ψ and ψ', and also the constants α, β, γ and δ. The potential ϕ at any point in (A+B) can then be computed using appropriately discretized forms of equations (33) and (34).

Solutions to Problem 1 have been obtained employing this modified BIE (MBIE) method, for the case OA = AB = EF = FO = 1 and $k_A = k_B = 1$, and are presented in Table 1d. Comparison with the solutions obtained employing the standard BIE methods, Table 1a, 1b and 1c, shows that the MBIE method affords a considerable improvement in the rate of convergence, in particular near the singularity. However, on the boundaries AB and DE, on which the potential is prescribed to be 0 and 1, respectively, the LBIE and QBIE methods are more accurate than the MBIE method.

Problem 2 has also been solved employing the MBIE method, and representative results for the case OA = AB = EF = FO = 1, h_1 = 1000, h_2 = 10, k_A = 250 and k_B = 10 are presented in Table 2d. These results are significantly better than those given by the standard BIE methods, Tables 2a, 2b and 2c. In particular, the heat transfer rates, Q_1 and Q_2, converge appreciably more rapidly and the requirement that the ratio of Q_1 and Q_2 be unity is satisfied more accurately than by the CBIE, LBIE and QBIE methods.

DISCUSSION AND CONCLUSIONS

The MBIE method presented here enables accurate solution of two-dimensional Laplacian problems involving singular points at which there is also a change of the physical properties. Although the method is only described for problems involving L-shaped domains, it is applicable to any such problem for which the analytical form of the singularity can be determined. Results have been obtained for other problems and in all cases the modified BIE method facilitated an improvement in the rate of convergence.

The additional sophistication inherent in the MBIE method, while requiring considerably more programming time than the standard BIE methods, affords improved accuracy for modest boundary discretizations. Furthermore, for a given number of boundary intervals, the MBIE only requires slightly more computational time than the CBIE and LBIE methods and in fact only requires between 1/4 and 1/2 the computational time of the QBIE method. This is due to the fact that for an N interval discretization, the QBIE method generates 2N x 2N equations, whereas the CBIE, LBIE and MBIE methods only generate N x N equations.

Although evaluation of the kernel integrals associated with the LBIE and QBIE methods, by the analytical expressions presented here, requires substantially less computational time than that required by previously employed quadrature formulae [13], these analytical expressions are only applicable for rectilinear boundaries; for the two problems considered in this study it has been found that the use of the analytical expressions facilitates a reduction in the overall computational time of up to 50 per cent, depending upon the piecewise-approximation and the size of the discretization. Thus, for problems involving curved boundaries it may be desirable to approximate the curved sections by a series of straight-line segments, and to integrate over these segments exactly.

It should be noted that the MBIE method is a modification of the CBIE method. The LBIE and QBIE methods cannot be modified, in a non-trivial way, because of the necessity to evaluate f_A' and f_B' at the point O, where these quantities are infinite. Further work on this aspect is at present under investigation.

REFERENCES

1. G.D. Smith, Numerical Solution of Partial Differential Equations , Oxford University Press, 1974.

2. O.C. Zienkiewicz, The Finite Element Method in Engineering , McGraw-Hill, London, 1971.

3. M.A. Jaswon and G.T. Symm, Integral Equation Methods in Potential Theory and Electrostatics , Academic Press, London, 1977.

4. P. Daly, "Singularities in Transmission Lines", The Mathematics of Finite Elements and Applications, Brunel University, 1972.

5. J.R. Whiteman, "Numerical Solution of Steady-State Diffusion Problems Containing Singularities", Proceedings of the International Symposium on Finite Elements in Flow Problems, Wiley, London, 1974.

6. G.T. Symm, "Treatment of Singularities in the Solution of Laplace's Equation by an Integral Equation Method", National Physical Laboratory, Report NAC31, 1973.

7. N. Papamichael and G.T. Symm, "Numerical Techniques for Two-Dimensional Laplacian Problems", Computer Methods in Applied Mechanics and Engineering, Vol. 6, 1975, pp. 175-194.

8. J.R. Whiteman and J.C. Webb, "Convergence of Finite-Difference Techniques for a Harmonic Mixed Boundary Value Problem", BIT, Vol. 10, 1970, pp. 336-374.

9. R. Wait, "Singular Isoparametric Finite Elements", The Journal of the Institute of Mathematics and Its Applications, Vol. 20, 1977, pp. 133-141.

10. D.F. Griffiths, "A Numerical Study of a Singular Elliptic Boundary Value Problem", The Journal of the Institute of Mathematics and Its Applications, Vol. 19, 1977, pp. 59-69.

11. P.J. Heggs and P.R. Stones, "The Effects of Dimensions on the Heat Flow Rate through Extended Surfaces", Journal of Heat Transfer, Vol. 102, 1980, pp. 180-182.

12. R.F. Harrington, K. Pontoppidan, P. Abrahamsen and N.C. Albertsen, "Computation of Laplacian Potentials by an Equivalent Source Method", Proceedings of the IEEE, Vol. 116, 1969, pp. 1715-1719.

13. G. Fairweather, F.J. Rizzo, D.J. Shippy and Y.S. Wu, "On the Numerical Solution of Two-Dimensional Potential Problems by an Improved Boundary Integral Equation Method", Journal of Computational Physics, Vol. 31, 1979, pp. 96-112.

Table 1a Problem 1 : CBIE Method Results

1.0064 1.0032 1.0016	0.9176 0.9170 0.9167	0.8310 0.8312 0.8313	0.7409 0.7419 0.7423	0.6458 0.6477 0.6484	0.5404 0.5453 0.5475
1.0009 0.9999 1.0000	0.9176 0.9176 0.9175	0.8326 0.8329 0.8331	0.7436 0.7445 0.7448	0.6497 0.6509 0.6513	0.5495 0.5518 0.5520
0.9998 1.0000 1.0000	0.9202 0.9204 0.9204	0.8382 0.8386 0.8387	0.7517 0.7524 0.7526	0.6588 0.6599 0.6602	0.5587 0.5598 0.5602
0.9999 1.0000 1.0000	0.9252 0.9254 0.9254	0.8482 0.8485 0.8486	0.7662 0.7668 0.7670	0.6756 0.6767 0.6770	0.5734 0.5749 0.5754
0.9999 1.0000 1.0000	0.9324 0.9325 0.9325	0.8630 0.8632 0.8633	0.7892 0.7896 0.7897	0.7050 0.7060 0.7064	0.5931 0.6006 0.6015
1.0000 1.0000 1.0000	0.9411 0.9411 0.9412	0.8818 0.8818 0.8818	0.8210 0.8210 0.8210	0.7571 0.7566 0.7565	0.6845 0.6785 0.6744
1.0000 1.0000 1.0000	0.9504 0.9503 0.9503	0.9017 0.9016 0.9015	0.8557 0.8554 0.8553	0.8164 0.8158 0.8155	0.7982 0.7969 0.7964
1.0001 1.0000 1.0000	0.9587 0.9586 0.9585	0.9194 0.9192 0.9191	0.8848 0.8845 0.8844	0.8594 0.8589 0.8588	0.8499 0.8492 0.8489
1.0000 1.0000 1.0000	0.9651 0.9649 0.9649	0.9326 0.9323 0.9322	0.9054 0.9050 0.9048	0.8867 0.8863 0.8861	0.8802 0.8796 0.8794
1.0004 1.0000 1.0000	0.9692 0.9689 0.9688	0.9406 0.9402 0.9401	0.9173 0.9169 0.9167	0.9019 0.9014 0.9012	0.8966 0.8960 0.8958
1.0024 1.0012 1.0006	0.9708 0.9703 0.9701	0.9433 0.9429 0.9427	0.9212 0.9207 0.9206	0.9067 0.9062 0.9060	0.9018 0.9012 0.9010

0.5562 0.5532 0.5515	0.4458 0.4456 0.4455	0.3364 0.3369 0.3371	0.2246 0.2256 0.2259	0.1110 0.1125 0.1131	-0.0087 -0.0044 -0.0022
0.5520 0.5517 0.5520	0.4466 0.4472 0.4474	0.3374 0.3382 0.3384	0.2256 0.2264 0.2267	0.1126 0.1133 0.1136	-0.0013 0.0001 0.0000
0.5587 0.5599 0.5602	0.4521 0.4531 0.4535	0.3410 0.3420 0.3423	0.2278 0.2287 0.2290	0.1139 0.1145 0.1147	0.0001 0.0000 0.0000
0.5733 0.5749 0.5754	0.4618 0.4634 0.4639	0.3465 0.3479 0.3483	0.2306 0.2317 0.2321	0.1151 0.1157 0.1160	0.0001 0.0000 0.0000
0.6026 0.6006 0.6015	0.4743 0.4768 0.4776	0.3522 0.3540 0.3546	0.2330 0.2345 0.2349	0.1157 0.1167 0.1170	-0.0013 0.0001 0.0000
0.6305 0.6430 0.6514	0.4820 0.4853 0.4864	0.3550 0.3570 0.3576	0.2338 0.2356 0.2361	0.1148 0.1167 0.1173	-0.0089 -0.0045 -0.0023

N = 50
100
200

Table 1b Problem 1 : LBIE Method Results

	D								C							B
	1.0000	0.9164	0.8310	0.7422	0.6485	0.5492		0.5492	0.4456	0.3375	0.2264	0.1135	0.0000
	1.0000	0.9165	0.8311	0.7422	0.6484	0.5491		0.5491	0.4451	0.3370	0.2260	0.1133	0.0000
	1.0000	0.9165	0.8312	0.7423	0.6485	0.5493		0.5493	0.4452	0.3369	0.2260	0.1133	0.0000
	1.0000	0.9173	0.8328	0.7445	0.6512	0.5519		0.5519	0.4478	0.3390	0.2273	0.1140	0.0000
	1.0000	0.9174	0.8329	0.7447	0.6512	0.5519		0.5519	0.4473	0.3384	0.2268	0.1137	0.0000
	1.0000	0.9175	0.8330	0.7448	0.6513	0.5520		0.5520	0.4473	0.3384	0.2268	0.1137	0.0000
	1.0000	0.9201	0.8384	0.7524	0.6601	0.5605		0.5605	0.4547	0.3435	0.2298	0.1151	0.0000
	1.0000	0.9203	0.8386	0.7525	0.6601	0.5602		0.5602	0.4537	0.3426	0.2292	0.1148	0.0000
	1.0000	0.9204	0.8387	0.7527	0.6602	0.5602		0.5602	0.4535	0.3424	0.2291	0.1147	0.0000
	1.0000	0.9251	0.8482	0.7666	0.6766	0.5762		0.5762	0.4666	0.3502	0.2332	0.1165	0.0000
	1.0000	0.9253	0.8485	0.7669	0.6769	0.5756		0.5756	0.4646	0.3488	0.2324	0.1161	0.0000
	1.0000	0.9254	0.8486	0.7671	0.6771	0.5755		0.5755	0.4642	0.3485	0.2322	0.1160	0.0000
	1.0000	0.9322	0.8627	0.7890	0.7056	0.6033		0.6033	0.4824	0.3571	0.2363	0.1177	0.0000
	1.0000	0.9324	0.8631	0.7896	0.7061	0.6023		0.6023	0.4791	0.3553	0.2353	0.1172	0.0000
	1.0000	0.9325	0.8633	0.7898	0.7065	0.6020		0.6020	0.4782	0.3549	0.2351	0.1172	0.0000
	1.0000	0.9408	0.8811	0.8199	0.7546	0.6658		0.6658	0.4897	0.3593	0.2370	0.1178	0.0000
	1.0000	0.9411	0.8817	0.8208	0.7562	0.6661		0.6661	0.4873	0.3580	0.2363	0.1175	0.0000
	1.0000	0.9412	0.8818	0.8210	0.7566	0.6664		0.6664	0.4868	0.3577	0.2362	0.1176	0.0000
	1.0000	0.9500	0.9008	0.8540	0.8132	0.7959	O						A
	1.0000	0.9503	0.9014	0.8551	0.8152	0.7968							
	1.0000	0.9503	0.9016	0.8553	0.8156	0.7966							
	1.0000	0.9583	0.9185	0.8833	0.8573	0.8484							
	1.0000	0.9585	0.9190	0.8843	0.8586	0.8490							
	1.0000	0.9585	0.9191	0.8844	0.8588	0.8490							
	1.0000	0.9647	0.9318	0.9041	0.8853	0.8791							
	1.0000	0.9649	0.9322	0.9048	0.8861	0.8795							
	1.0000	0.9649	0.9322	0.9049	0.8862	0.8795							
	1.0000	0.9687	0.9398	0.9162	0.9007	0.8957							
	1.0000	0.9688	0.9401	0.9167	0.9012	0.8959							
	1.0000	0.9688	0.9401	0.9167	0.9013	0.8958							
	1.0000	0.9699	0.9423	0.9200	0.9053	0.9007							
	1.0000	0.9701	0.9427	0.9205	0.9060	0.9011							
	1.0000	0.9701	0.9427	0.9206	0.9061	0.9010							
	E						F							

N = 50, 100, 200

Table 1c   Problem 1 : QBIE Method Results

D											C
1.0000 1.0000 1.0000	0.9165 0.9165 0.9165	0.8311 0.8311 0.8312	0.7420 0.7422 0.7423	0.6480 0.6483 0.6485	0.5484 0.5490 0.5493	0.5484 0.5490 0.5493	0.4442 0.4448 0.4451	0.3361 0.3366 0.3369	0.2253 0.2257 0.2259	0.1129 0.1132 0.1133	0.0000 0.0000 0.0000
1.0000 1.0000 1.0000	0.9174 0.9175 0.9175	0.8329 0.8330 0.8330	0.7445 0.7447 0.7448	0.6508 0.6511 0.6513	0.5512 0.5516 0.5519	0.5512 0.5516 0.5519	0.4464 0.4469 0.4472	0.3375 0.3380 0.3383	0.2261 0.2265 0.2267	0.1134 0.1136 0.1137	0.0000 0.0000 0.0000
1.0000 1.0000 1.0000	0.9204 0.9204 0.9204	0.8386 0.8387 0.8387	0.7524 0.7526 0.7527	0.6598 0.6600 0.6602	0.5594 0.5599 0.5601	0.5594 0.5599 0.5601	0.4526 0.4531 0.4534	0.3415 0.3420 0.3423	0.2284 0.2287 0.2289	0.1143 0.1145 0.1147	0.0000 0.0000 0.0000
1.0000 1.0000 1.0000	0.9254 0.9254 0.9254	0.8487 0.8487 0.8487	0.7670 0.7671 0.7671	0.6768 0.6770 0.6771	0.5747 0.5751 0.5754	0.5747 0.5751 0.5754	0.4629 0.4635 0.4639	0.3474 0.3479 0.3483	0.2314 0.2318 0.2320	0.1156 0.1158 0.1159	0.0000 0.0000 0.0000
1.0000 1.0000 1.0000	0.9326 0.9326 0.9325	0.8635 0.8634 0.8634	0.7901 0.7900 0.7899	0.7067 0.7066 0.7066	0.6011 0.6014 0.6016	0.6011 0.6014 0.6016	0.4762 0.4770 0.4776	0.3533 0.3541 0.3546	0.2340 0.2346 0.2349	0.1166 0.1169 0.1171	0.0000 0.0000 0.0000
1.0000 1.0000 1.0000	0.9414 0.9413 0.9412	0.8822 0.8820 0.8819	0.8218 0.8215 0.8212	0.7579 0.7573 0.7569	0.6659 0.6663 0.6665	0.6659 0.6663 0.6665	0.4837 0.4852 0.4861	0.3557 0.3568 0.3574	0.2349 0.2357 0.2361	0.1168 0.1173 0.1175	0.0000 0.0000 0.0000
1.0000 1.0000 1.0000	0.9506 0.9504 0.9504	0.9022 0.9019 0.9017	0.8563 0.8559 0.8555	0.8173 0.8165 0.8159	0.7993 0.7979 0.7969						O
1.0000 1.0000 1.0000	0.9589 0.9587 0.9586	0.9198 0.9194 0.9192	0.8854 0.8849 0.8846	0.8603 0.8595 0.8591	0.8508 0.8498 0.8492						
1.0000 1.0000 1.0000	0.9652 0.9650 0.9649	0.9328 0.9325 0.9323	0.9058 0.9053 0.9050	0.8874 0.8868 0.8864	0.8809 0.8801 0.8796						
1.0000 1.0000 1.0000	0.9691 0.9689 0.9688	0.9407 0.9404 0.9402	0.9176 0.9171 0.9168	0.9024 0.9018 0.9014	0.8971 0.8964 0.8960						
1.0000 1.0000 1.0000	0.9704 0.9702 0.9701	0.9433 0.9430 0.9428	0.9215 0.9210 0.9207	0.9072 0.9066 0.9062	0.9023 0.9016 0.9012						
E					F						A/B

$N = 50, 100, 200$

Table 1d   Problem 1 : MBIE Method Results

D
1.0005	0.9167	0.8312	0.7423	0.6486	0.5495						
1.0001	0.9166	0.8313	0.7424	0.6487	0.5496						
1.0000	0.9166	0.8313	0.7424	0.6487	0.5496						
0.9999	0.9175	0.8331	0.7448	0.6514	0.5521	0.5494	0.4453	0.3371	0.2261	0.1135	0.0001
1.0000	0.9175	0.8331	0.7449	0.6515	0.5521	0.5495	0.4453	0.3371	0.2261	0.1134	0.0000
1.0000	0.9175	0.8331	0.7449	0.6515	0.5522	0.5496	0.4454	0.3371	0.2261	0.1134	0.0000
1.0000	0.9204	0.8387	0.7527	0.6603	0.5603	0.5521	0.4473	0.3384	0.2269	0.1138	0.0000
1.0000	0.9204	0.8387	0.7527	0.6603	0.5604	0.5521	0.4474	0.3385	0.2269	0.1138	0.0000
1.0000	0.9204	0.8388	0.7528	0.6604	0.5604	0.5522	0.4474	0.3385	0.2269	0.1138	0.0000
1.0000	0.9254	0.8487	0.7671	0.6772	0.5756	0.5603	0.4536	0.3425	0.2291	0.1147	0.0000
1.0000	0.9254	0.8487	0.7671	0.6772	0.5756	0.5604	0.4536	0.3425	0.2291	0.1147	0.0000
1.0000	0.9254	0.8487	0.7671	0.6772	0.5756	0.5604	0.4536	0.3425	0.2291	0.1147	0.0000
1.0000	0.9325	0.8633	0.7898	0.7066	0.6019	0.5756	0.4641	0.3485	0.2322	0.1160	0.0000
1.0000	0.9325	0.8633	0.7898	0.7066	0.6019	0.5756	0.4642	0.3486	0.2323	0.1161	0.0000
1.0001	0.9325	0.8633	0.7898	0.7066	0.6019	0.5756	0.4642	0.3486	0.2323	0.1161	0.0000
1.0000	0.9412	0.8818	0.8210	0.7565	0.6667	0.6019	0.4780	0.3549	0.2352	0.1172	0.0000
1.0000	0.9412	0.8818	0.8211	0.7565	0.6667	0.6019	0.4780	0.3549	0.2352	0.1172	0.0000
1.0001	0.9412	0.8818	0.8211	0.7565	0.6667	0.6019	0.4780	0.3549	0.2352	0.1172	0.0000
1.0000	0.9503	0.9016	0.8553	0.8154	0.7962	0.6667	0.4869	0.3579	0.2364	0.1177	0.0000
1.0001	0.9503	0.9016	0.8553	0.8154	0.7961	0.6667	0.4869	0.3579	0.2364	0.1177	0.0000
1.0001	0.9503	0.9016	0.8553	0.8154	0.7961	0.6667	0.4869	0.3579	0.2364	0.1177	0.0000
1.0001	0.9586	0.9191	0.8844	0.8587	0.8488					A	
1.0001	0.9586	0.9191	0.8844	0.8587	0.8487						
1.0001	0.9586	0.9191	0.8844	0.8587	0.8487						
1.0001	0.9649	0.9322	0.9049	0.8862	0.8794						
1.0001	0.9649	0.9322	0.9049	0.8862	0.8793						
1.0001	0.9649	0.9322	0.9049	0.8862	0.8793						
1.0002	0.9687	0.9401	0.9167	0.9013	0.8958						
1.0001	0.9688	0.9401	0.9167	0.9013	0.8957						
1.0001	0.9688	0.9401	0.9168	0.9014	0.8957						
0.9995	0.9699	0.9427	0.9207	0.9062	0.9010						
1.0001	0.9701	0.9428	0.9207	0.9062	0.9009						
1.0002	0.9702	0.9428	0.9207	0.9063	0.9009						
E   F   O   C   B

N = 50, 100, 200

Table 2a   Problem 2:   CBIE Method Results

	Intervals		
	50	100	200
Q_1	5.7387	5.7346	5.7330
Q_2	5.7342	5.7321	5.7317
Q_1/Q_2	1.0007	1.0004	1.0002

Table 2b   Problem 2:   LBIE Method Results

	Intervals		
	50	100	200
Q_1	5.7084	5.7199	5.7260
Q_2	5.7387	5.7350	5.7334
Q_1/Q_2	0.9947	0.9973	0.9987

Table 2c   Problem 2:   QBIE Method Results

	Intervals		
	50	100	200
Q_1	5.7129	5.7231	5.7280
Q_2	5.7334	5.7325	5.7321
Q_1/Q_2	0.9964	0.9984	0.9993

Table 2d   Problem 2: MBIE Method Results

	Intervals		
	50	100	200
Q_1	5.7325	5.7321	5.7321
Q_2	5.7330	5.7325	5.7321
Q_1/Q_2	0.9999	1.0000	1.0000

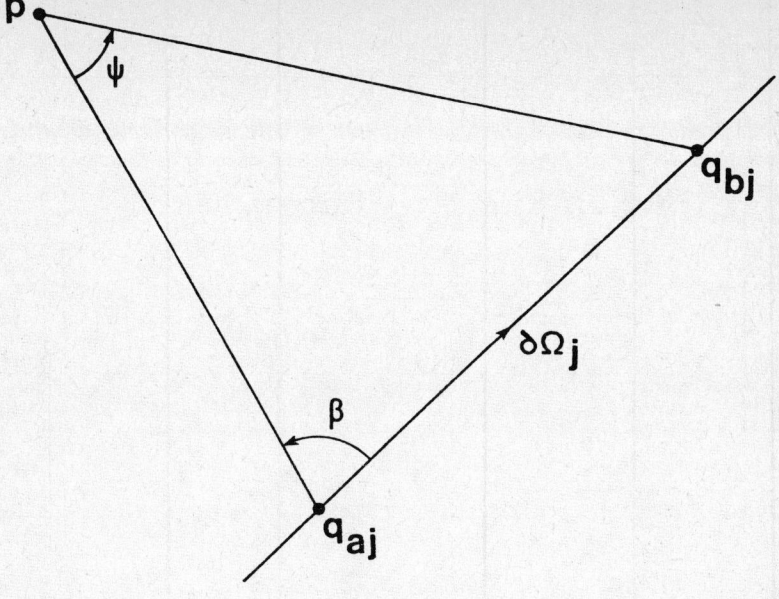

Fig.1 Straight-line segment geometry

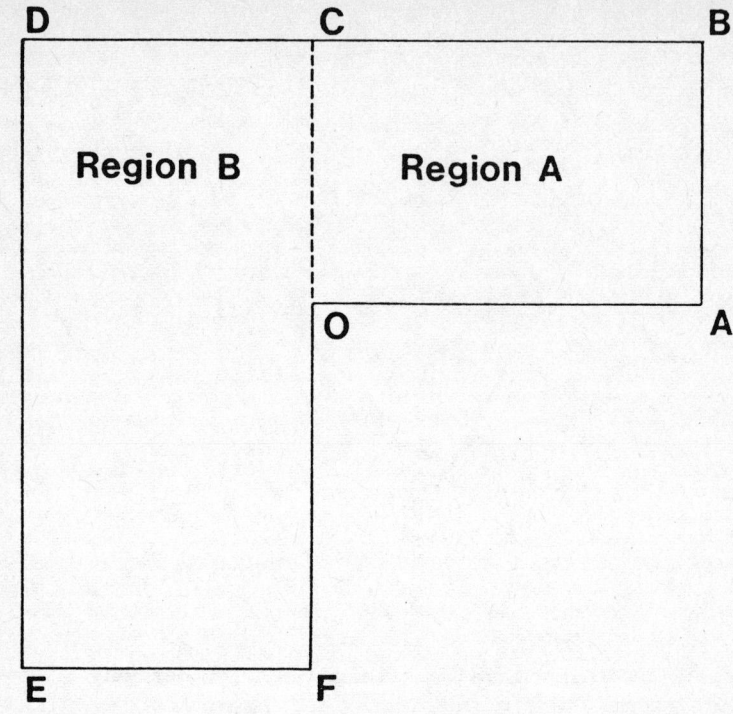

Fig.2 L-shaped solution domain

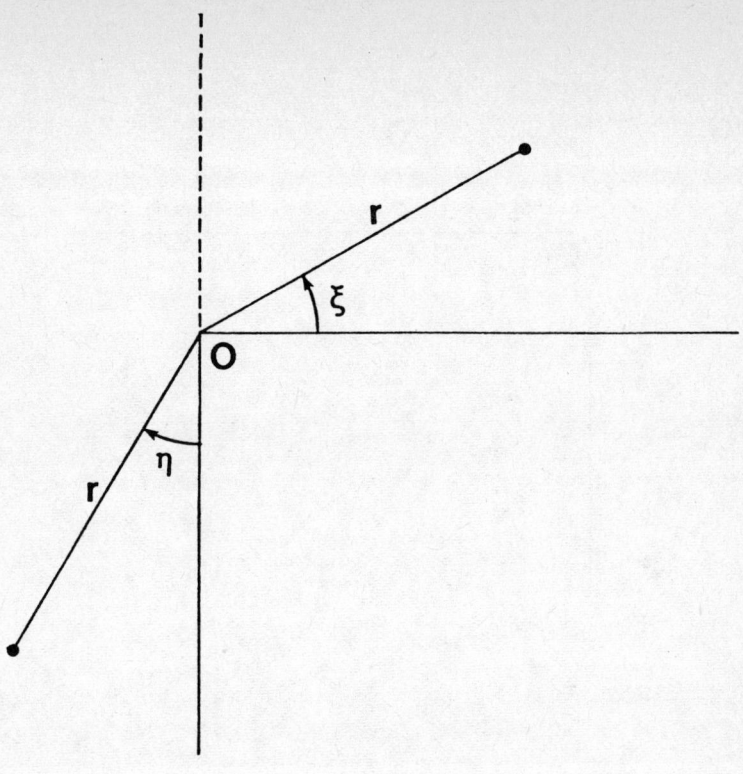

Fig.3 Neighbourhood of the singularity

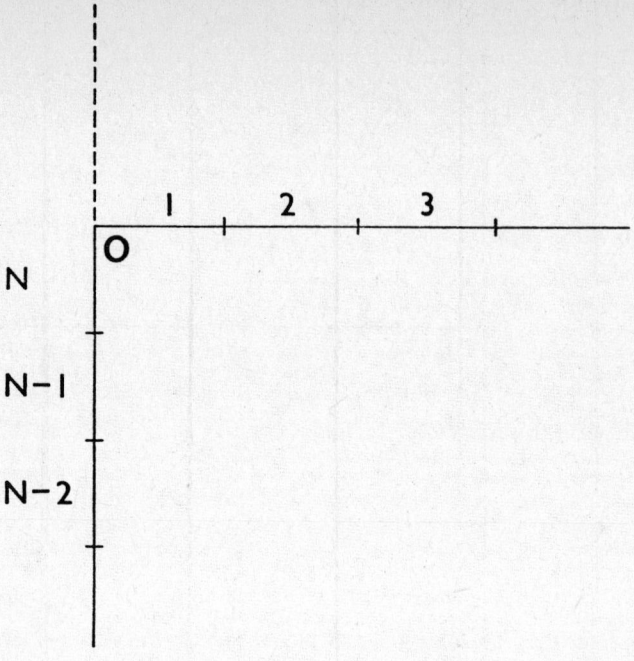

Fig.4 Boundary discretization in the neighbourhood of the singularity

3.2 THE BOUNDARY INTEGRAL EQUATION ANALYSIS OF TRANSMISSION LINE SINGULARITIES

ABSTRACT

The transverse electromagnetic line analysis of microstrips and co-axial lines generally involves boundary singularities which cause slow convergence of the solutions computed by standard numerical techniques. In this study the boundary singularities occurring at the ends of the inner conductor in an unsymmetric closed microstrip containing a dielectric substrate are treated by a modified boundary integral equation method. This method is shown to be successful in reducing the error due to the presence of the singularities.

INTRODUCTION

The transverse electromagnetic line analysis of microstrips and co-axial lines generally involves field singularities due either to re-entrant corners or discontinuous boundary conditions, e.g. [1-4]. The presence of these boundary singularities affects the accuracy of the solutions computed by the standard forms of numerical techniques such as the boundary integral equation (BIE), [5], finite-difference [6] and finite-element [7] methods. Consequently, the possibility of modifying these numerical techniques to give special treatment to the singular points and thereby to obtain solutions which converge more rapidly has received considerable attention, e.g. [1-4, 8-16].

Modified finite-difference formulations employ mesh refinement or special difference equations in the neighbourhood of the singular points, [11,12,13], whilst advanced finite-element implementations involve the use of either higher order elements combined with mesh refinement or singular shape functions in the elements near the singularity, [14,15,16]. Local mesh refinement is of limited value because the primary cause of the inaccuracy in numerical solutions is the presence of the boundary singularity, not the discretization error. Modified finite-difference and finite-element formulations which account for the analytical behaviour of the singularity usually facilitate improved accuracy, e.g. [1,2,3], but suffer from the disadvantage that a systematic treatment of all the nodel points is not possible and consequently necessitate a considerably more complicated program than required for the standard implementations. However, Symm [8] has devised a modified BIE method which can successfully treat boundary singularities in plane Laplacian problems, but without any appreciable increases in either the programming complexity or the computational requirements. The results obtained employing this technique offer a considerable improvement over those given by modified finite-difference and finite-element methods, [8,9,10]. Furthermore, the BIE method has the advantage that the discretization for numerical purposes occurs only on the boundary of the relevant domain and therefore generates a considerably smaller system of equations than an equivalent finite-difference or finite-element representation.

Ingham et al [10] have recently extended the modified BIE method presented by Symm [8] to enable the accurate analysis of problems in which the boundary singularity occurs on the interface of regions which have different physical properties. In this study the technique described by Ingham et al [10] is employed in order to determine the the capacitance of an unsymmetric closed microstrip line containing a dielectric substrate, Fig. 1, for which a singularity occurs at each of the ends of the inner conductor OO', Fig. 1, on the interface between air and dielectric. The solutions obtained employing the modified BIE method are contrasted with those given by the standard BIE method and the accuracy of the BIE solutions is illustrated by considering a symmetric problem for which the exact solution is known.

APPLICATION OF THE BIE METHODS

Consider a microstrip consisting of a homogeneous dielectric substrate with constant permittivity, κ, enclosed within a rectangular case, as shown schematically in Fig. 1. The geometrical symmetry of the microstrip configuration indicates that it is only necessary to examine the region ABCDEFA, Fig. 1. Thus, the field distribution ϕ within the microstrip satisfies,

$$\nabla^2 \phi_A = 0 \quad \text{in region A, Fig. 1,} \tag{1}$$

and
$$\nabla^2 \phi_B = 0 \quad \text{in region B, Fig. 1,} \tag{2}$$

subject to the boundary conditions,

on OA	$\phi_A = 1$	(3a)
on AB	$\phi_A' = 0$	(3b)
on BC	$\phi_A = 0$	(3c)
on CD	$\phi_A = 0$	(3d)
on DO	$\phi_A = \phi_B$	(3e)
and	$\phi_A' = -\kappa \phi_B'$	(3f)
on DE	$\phi_B = 0$	(3g)

on EF $\quad \phi_B = 0$ (3h)

on FA $\quad \phi_B' = 0$ (3i)

on AO $\quad \phi_B = 1$ (3j)

where the potential is non-dimensionalised such that the inner and outer conductors have unit and zero potential respectively and ϕ' denotes the normal derivative of the potential. Conditions (3b) and (3i) arise from the symmetry of the microstrip cross-section about the axis BAF, and conditions (3e) and (3f) stipulate continuity of the electric field and normal intensity across the air-dielectric interface DO.

This problem has a singularity at the re-entrant corner O, in the neighbourhood of which the potential is described by [2],

$$\phi_A(r,\xi) = 1 + \alpha r^{\frac{1}{2}} \sin \frac{1}{2}\xi + \beta r \sin\xi + \gamma r^{3/2} \sin \frac{3}{2}\xi + \delta r^2 \sin 2\xi + \ldots$$

in region A, Fig. 1 (4)

and

$$\phi_B(r,\eta) = 1 + \alpha r^{\frac{1}{2}} \sin \frac{1}{2}\eta + (-\frac{\beta}{\kappa}) r \sin\eta + \gamma r^{3/2} \sin \frac{3}{2}\eta + (\frac{-\delta}{\kappa}) r^2 \sin 2\xi + \ldots$$

in region B, Fig. 1 (5)

where (r,ξ) and (r,η) are polar coordinates as defined in Fig. 1. From the expressions (4) and (5) it is apparent that all derivatives of the potential become unbounded as the singular point is approached.

In order to illustrate the problems caused by the presence of the singularity, solutions are first obtained employing the standard BIE method, i.e. without giving any special attention to the singularity. Applying Green's Integral Formula [5], the problem described by equations (1), (2) and (3) is reformulated in the form of two coupled integral equations involving contour integrals around OABCDO and ODEFAO; the coupling arises from the interface conditions (3e) and (3f). In order to effect a solution to these integral equations, the contours OABCDO and ODEFAO are subdivided into a total of N rectilinear segments and on each of these segments the potential ϕ and its normal derivative ϕ' are approximated by piecewise-constant functions. Then, Green's Boundary Formula [5] is applied successively to the midpoint of each boundary segment, generating a system of linear algebraic equations in which the unknowns are the boundary values of ϕ and ϕ' complementary to those prescribed by the boundary conditions (3). Solution of this system of equations is achieved by a Gaussian elimination technique [17] and determines the boundary distributions of both the potential and its normal derivative.

The capacitance per unit length of the microstrip is given by,

$$C = 2\varepsilon_o \{\int_{OA} \phi_A'(q) dq + \int_{AO} \kappa \, \phi_B'(q) dq\}, \qquad (6)$$

where ε_o is the free-space permittivity. Expression (6) only involves the boundary distribution of the normal intensity and can therefore be evaluated directly from the BIE boundary solution. Thus, unlike the finite-difference and finite-element methods, the BIE method affords the advantage that the microstrip capacitance can be evaluated without computing the field distribution within the microstrip.

The standard BIE solutions for the capacitance of three particular microstrip configurations are presented in Table 1. For each of the three problems solutions were obtained employing 60, 120, 240 and 480 boundary segments, corresponding to 5, 10, 20 and 40 segments on each of the sides OA,AB,BX,XC,CD,DO,OD,DE,EY,YF,FA and AO Fig. 1.

The slow convergence caused by the presence of the boundary singularity at the corner O is clearly illustrated by these results.

The modified BIE method [8,9,10] attempts to reduce the singularity error by incorporating the analytical behaviour of the singularity. This is accomplished by introducing a new potential ψ such that,

$$\phi_A = \psi_A + f_A \quad \text{in} \quad \text{region A, Fig. 1,} \tag{7}$$

and
$$\phi_B = \psi_B + f_B \quad \text{in} \quad \text{region B, Fig. 1,} \tag{8}$$

where f_A and f_B are the singular functions,

$$f_A(r,\xi) = 1 + \alpha r^{1/2} \sin \frac{1}{2}\xi + \beta r \sin \xi + \gamma r^{3/2} \sin \frac{3}{2}\xi + \delta r^2 \sin 2\xi \tag{9}$$

and

$$f_B(r,\eta) = 1 + \alpha r^{1/2} \sin \frac{1}{2}\eta + \left(\frac{-\beta}{\kappa}\right) r \sin \eta + \gamma r^{3/2} \sin \frac{3}{2}\eta + \left(\frac{-\delta}{\kappa}\right) r^2 \sin 2\xi \tag{10}$$

extracted from the equations (4) and (5).

Thus, the potential ψ satisfies the boundary-value problem,

$$\nabla^2 \psi_A = 0 \quad \text{in} \quad \text{region A, Fig. 1,} \tag{11}$$

and
$$\nabla^2 \psi_B = 0 \quad \text{in} \quad \text{region B, Fig. 1,} \tag{12}$$

subject to,

on OA $\quad \psi_A = 1 - f_A$ (13a)

on AB $\quad \psi_A' = -f_A'$ (13b)

on BC $\quad \psi_A = -f_A$ (13c)

on CD $\quad \psi_A = -f_A$ (13d)

on DO $\quad \psi_A = \psi_B$ (13e)

and $\quad \psi_A' = -\kappa \psi_B'$ (13f)

on DE $\quad \psi_B = -f_B$ (13g)

on EF $\quad \psi_B = -f_B$ (13h)

on FA $\quad \psi_B' = -f_B'$ (13i)

on AO $\quad \psi_B = 1 - f_B$ (13j)

The application of the standard BIE method to the problem described by equations (11), (12) and (13) generates a system of N linear algebraic equations in N + 4 unknowns, including the constants α, β, γ and δ. In order to reduce the number of unknowns to N it is necessary to assume that $\psi_A = 0$ on the segments 1 and N, Fig. 1, and $\psi_B = 0$ on the segments $\frac{N}{2}$ and $\frac{N}{2} + 1$, Fig. 1, i.e. in the vicinity of the singular point O the potential ϕ can be approximated by the expressions (9) and (10) for f_A and f_B. Solving the modified system of equations determines the boundary distributions of ψ and ψ', and also the constants α, β, γ and δ. The potential ϕ may then be computed employing the relations (7) and (8).

The results predicted by the modified BIE method, for the capacitance of the three microstrip configurations considered earlier, are presented in Table 2. These results were obtained employing exactly the same distribution of boundary segments as used to obtain the standard BIE method solutions. Comparison with the solutions given by the standard BIE method, Table 1, shows that the modified method affords a considerable improvement in the rate of convergence; in all cases the first three significant figures of the solutions predicted by the modified BIE method remain unchanged as the boundary discretization is refined from 120 to 480 segments, Table 2.

Problem (1), in Tables 1 and 2, correspond to the case of a centre conductor in a stripline. For this problem the exact solution can be determined by conformal transformation techniques and is known to be $C/\varepsilon_o = 5.87688$, [18]. The error in the standard BIE method solution is greater than 0.3 per cent for the finest discretization, i.e. N = 480. In contrast, the modified BIE method solutions differ from the exact solution by only 0.04 percent for the 60 segment discretization and by less than 0.002 percent for the 480 segment discretization.

Problem (2), in Tables 1 and 2, represent an unsymmetric closed microstrip without a dielectric and has previously been treated by a modified finite-element method involving mesh refinement near the singularity, [4]. The capacitance predicted by this modified finite-element method is $C/\varepsilon_o = 4.77708$ and differs quite significantly from that given by the modified BIE method, Table 2. However, whereas the BIE solutions are obtained by a uniform refinement of all boundary segments in the ratio $1 : \frac{1}{2} : \frac{1}{4} : \frac{1}{8}$, the modified finite-element method results were obtained by establishing a mesh over the domain of the microstrip and then subdividing the mesh only in the vicinity of the singularity. This procedure may nullify the effect of the singularity and thereby produce more rapid convergence, but has the disadvantage that it contains a discretization error which, at points remote from the singularity, is related to the original mesh size. Thus, the apparent convergence will not be to the correct limit.

Problem (3), in Tables 1 and 2, corresponds to an unsymmetric closed microstrip of the same dimensions as that considered in problem (2), but including a dielectric with permittivity $\kappa = 10$. Comparison of the results for problems (2) and (3) shows that the introduction of the dielectric does not affect the rate of convergence of the modified BIE method solutions.

DISCUSSION AND CONCLUSIONS

The modified BIE method [10] has been shown to give accurate results for a transmission line problem in which boundary singularities occur on the interface between regions with different permittivity. This BIE method has several computational advantages over comparable finite-difference and finite-element methods. The principal advantage is that, in contrast with the finite-difference and finite-element methods, the BIE discretization occurs only on the microstrip boundary and therefore generates a considerably smaller system of algebraic equations than an equivalent finite-difference of finite-element approximation. Thus, the BIE formulation facilitates substantial reductions in the computational storage and time requirements in comparison with equivalent finite-difference and finite-element representations. In addition, the BIE formulation has the inherent feature that the microstrip capacitance can be calculated without determining the field distribution within the microstrip. This is due to the fact that in the BIE discretization the unknowns are the boundary values of ϕ and ϕ' complementary to those prescribed by the boundary conditions. The solution of the BIE numerical representation determines the boundary distribution of both ϕ and ϕ' and therefore provides all the information necessary to enable the computation of the capacitance. Furthermore, with the BIE technique the potential at points within the microstrip can be computed directly from the boundary distribution of ϕ and ϕ' and need be determined only if desired.

REFERENCES

1. J.W. Duncan, "The Accuracy of Finite Difference Solutions of Laplace's Equation", IEEE Transactions, Vol. MTT-15, pp. 575-582, 1967.

2. K.B. Whiting, "A Treatment for Boundary Singularities in Finite Difference Solutions of Laplace's Equation", IEEE Transactions, Vol. MTT-16, pp. 889-891, 1968.

3. J.E. Akin, "Finite Element Analysis of Fields with Boundary Singularities", International Conference on Numerical Methods in Electrical and Magnetic Field Problems, Santa Margherita, Italy, pp. 60-72, 1976.

4. P. Daly, "Singularities in Transmission Lines", The Mathematics of Finite Elements and Applications, edited by J.R. Whiteman, Academic, pp. 337-350, 1973.

5. M.A. Jaswon and G.T. Symm, Integral Equation Methods in Potential Theory and Electrostatics, Academic Press, London, 1977.

6. G.D. Smith, Numerical Solution of Partial Differential Equations, Oxford University Press, 1974.

7. O.C. Zienkiewicz, The Finite Element Method in Engineering, McGraw-Hill, London, 1971.

8. G.T. Symm, "Treatment of Singularities in the Solution of Laplace's Equation by an Integral Equation Method", National Physical Laboratory, Report NAC 31, 1973.

9. N. Papamichael and G.T. Symm, "Numerical Techniques for Two-Dimensional Laplacian Problems", Computer Methods in Applied Mechanics and Engineering, Vol. 6, pp 175-194, 1975.

10. D.B. Ingham, P.J. Heggs and M. Manzoor, "The Numerical Solution of Plane Potential Problems by Improved Boundary Integral Equation Methods, Submitted to Journal of Computational Physics, 1980.

11. H. Motz, "The Treatment of Singularities of Partial Differential Equations by Relaxation Methods", Quarterly of Applied Mathematics, Vol. 4, pp. 371-377, 1947.

12. A.F. Emery, "The Use of Singularity Programming in Finite Difference and Finite Element Computations of Temperature", Journal of Heat Transfer, Vol. 95, pp. 344-351, 1973.

13. J.R. Whiteman, "Numerical Solution of Steady-State Problems Containing Singularities", Finite Elements in Fluids, edited by R.H. Gallagher, Wiley, 1975.

14. R.W. Thatcher, "Singularities in the solution of Laplace's Equation in Two-Dimensions", Journal of the Institute of Mathematics and Its Applications, Vol. 16, pp. 303-319, 1975.

15. R.E. Barhill and J.R. Whiteman, "Error Analysis of Finite Element Methods with Triangles for Elliptic Boundary-Value Problems", The Mathematics of Finite Element and Applications, edited by J.R. Whiteman, Academic, pp. 83-112, 1973.

16. D.F. Griffiths, "A Numerical Study of a Singular Elliptic Boundary-Value Problem", Journal of the Institute of Mathematics and Its Applications, Vol. 19, pp. 56-69, 1977.

17. A. Ralston, A First Course in Numerical Analysis, McGraw-Hill, New York, 1965.

18. R. Levy, "New Coaxial-to-Stripline Transformers Using Rectangular Lines", IEEE Transactions, Vol. MTT-17, pp. 204-210, 1969.

Table 1 : Results Predicted by the Standard BIE Method

Problem		Capacitance, C/ε_o			
		N=60	N=120	N=240	N=480
(1)	a=2b=2w=4h, κ=1	5.76125	5.81228	5.84241	5.85896
(2)	3a=8b=12w=24h, κ=1	4.72859	4.76957	4.79272	4.80509
(3)	3a=8b=12w=24h, κ=10	30.06730	30.42114	30.60443	30.69767

Table 2 : Results Predicted by the Modified BIE Method

Problem		Capacitance, C/ε_o			
		N=60	N=120	N=240	N=480
(1)	a=2b=2w=4h, κ=1	5.87961	5.87776	5.87716	5.87696
(2)	3a=8b=12w=24h, κ=1	4.84464	4.82694	4.82098	4.81902
(3)	3a=8b=12w=24h, κ=10	31.03727	30.87552	30.81925	30.80047

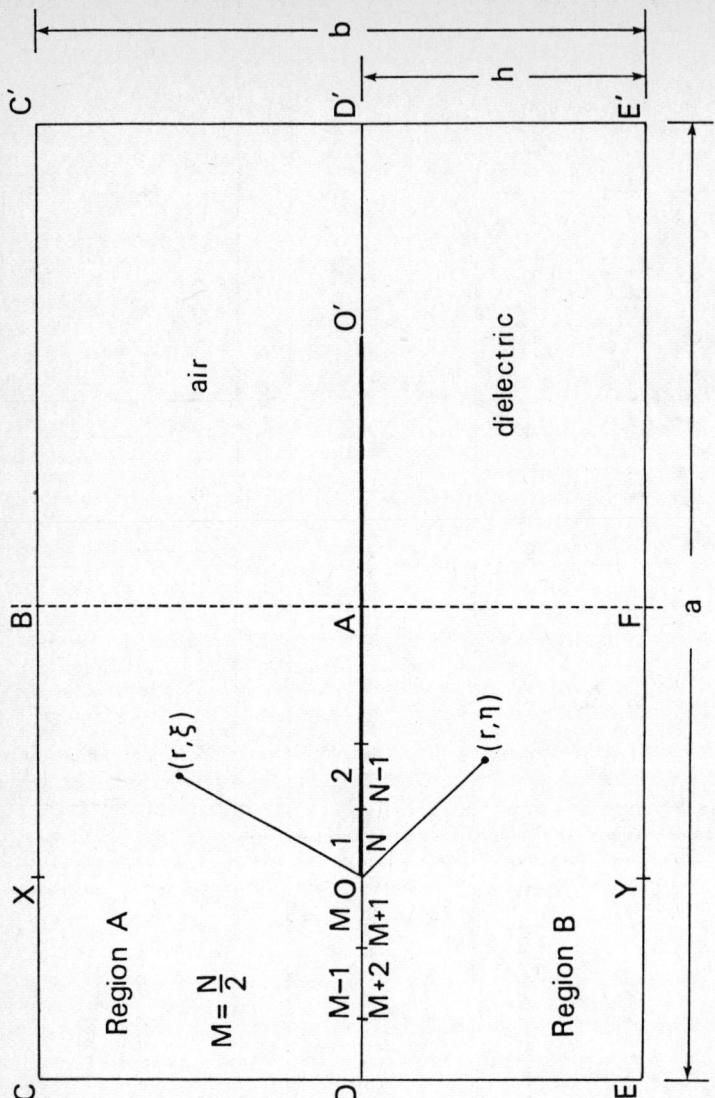

Fig.1 Schematic representation of an unsymmetric microstrip

3.3 THE BOUNDARY INTEGRAL EQUATION ANALYSIS OF FIN ASSEMBLY HEAT TRANSFER

ABSTRACT

In this study the boundary integral equation (BIE) formulation is developed for predicting the heat transfer in situations where heat is exchanged between two fluids which are separated by a finned interface. With this BIE formulation discretization for numerical purposes occurs only on the boundary of the domain used to represent the finned surface. Thus, in comparison with other equivalent numerical techniques, the BIE method facilitates a considerable reduction in the computational requirements. The flexibility and conceptual simplicity of the BIE method enables quite general linear boundary conditions to be accommodated without difficulty. In particular, the BIE method can easily handle problems in which the fins and the supporting surface have different thermal conductivities. It is shown in this study that the classical BIE method, which employs a relatively crude piecewise-constant approximation, gives an effective treatment even when comparable finite-difference and finite-element implementations fail to provide a satisfactory solution. The introduction of higher order BIE approximations necessitates an additional programming effort but facilities improved accuracy without any appreciable increase in the computational requirements. For problems involving small fin length to fin thickness ratios further improvements in the accuracy of the BIE solutuons can be achieved by providing special treatment to boundary singularities.

INTRODUCTION

The theoretical study of the heat flow within finned heat exchangers is of considerable practical importance because of the extensive utilisation of fins for heat transfer enhancement in applications varying from gas liquefaction plant to heat rejection equipment in motor vehicle engines. The accurate prediction of the thermal performance of finned heat exchangers is essential for compact and efficient design. However, the analysis of such systems is conventionally based on several simplifying assumptions, in particular, that the heat flow is one-dimensional. This assumption, in general, permits an analytical treatment, e.g. [1,2,3].

The early investigations into the applicability of the one-dimensional approximation restricted attention solely to the fin and concluded that two-dimensional effects are negligible provided the transverse Biot number, based on the fin thickness, is much less than unity [4,5,6]. However, recent investigations of the combined fin and supporting surface have shown that the presence of fins induces two-dimensional effects within the supporting surface and these may in turn act to produce two-dimensional variations within the fin [7-11]. Suryanarayana [8] has reported that the difference between the heat transfer rates predicted by one- and two-dimensional analyses can be as much as 80 per cent. It is therefore essential for the effective design of finned heat exchangers to consider the complete fin assembly and to employ a multi-dimensional analysis.

The two-dimensional analysis of conductive-convective heat flow through an assembly of longitudinal fins attached to a plane wall, Fig. 1, requires the solution of a Laplacian mixed boundary-value problem [7-11]. The combined complexity of the boundary conditions and the domain geometry precludes an analytical treatment. However, various numerical techniques may be employed in order to achieve a solution, e.g. the finite-difference [12], the finite-element [13] and the boundary integral equation [14] methods. In previous examinations of fin assembly heat transfer, Suryanarayana [8], Heggs and Stones [9,10] and Stones [11], have all employed the finite-difference method, whilst Stones [11] has also used a finite-element method. A comparison of the finite-difference and finite-element methods, performed by Stones [11], has revealed that for a particular range of the system parameters neither of these methods produces satisfactory solutions; the finite-difference solutions fail to satisfy the energy conservation criteria, whilst the finite-element solutions exhibit a temperature discontinuity at the wall-to-fin interface.

The boundary integral equation (BIE) method has not previously been applied to the fin assembly problem although it affords several computational advantages over the finite-difference and finite-element methods for the solution of Laplacian mixed boundary-value problems [14,15,16,17]. The principal advantage is that, in contrast to the finite-difference and finite-element methods, the BIE discretization occurs only on the boundary of the relevant domain and therefore generates a considerably smaller system of algebraic equations than an equivalent finite-difference or finite-element approximation. Thus, the BIE formulation facilitates a substantial reduction in the computational storage and time requirements. In addition, the flexibility of the BIE technique permits complexities, such as curved fin profiles, non-uniform variations of the surface heat transfer coefficients and different thermal conductivities of the fin and supporting surface materials, to be accommodated with ease. Furthermore, the BIE method has the inherent feature that the surface heat flux can be evaluated without computing the temperature distribution within the fin assembly.

The primary objective of the present study is to demonstrate the advanced solution capabilities of the BIE method for the analysis of two-dimensional heat conduction problems. This is achieved by investigating the performance of the BIE method in determining the temperature distribution and heat transfer rate of an assembly of longitudinal fins attached to a plane wall, Fig. 1.

ANALYSIS

The following analysis is based upon the classical assumptions employed in the examination of conducting-convecting finned surfaces, namely, constant thermal conductivities, uniform heat transfer coefficients and perfect wall-to-fin contact. These simplifications are introduced not only to reduce the complexity of the problem but also to minimise the number of variable parameters whilst still retaining the essential features of the actual physical situation.

The geometrical symmetry of an assembly of equally-spaced longitudinal rectangular fins attached to a plane wall indicates that it is only necessary to examine that section of the assembly shown schematically in Fig. 1. Thus, for steady-state two-dimensional heat flow, the determination of the assembly temperature distribution ϕ requires the simultaneous solution of [9],

$$\nabla^2 \phi_f = 0 \quad \text{within the fin,} \tag{1}$$

and
$$\nabla^2 \phi_w = 0 \quad \text{within the wall,} \tag{2}$$

subject to the boundary conditions

on OA $\quad \phi_f' = -\dfrac{Bi_2}{\kappa} \phi_f \tag{3a}$

on AB $\quad \phi_f' = -\dfrac{Bi_2}{\kappa} \phi_f \tag{3b}$

on BC $\quad \phi_f' = 0 \tag{3c}$

on CO $\quad \phi_f = \phi_w$ (3d)

and $\quad \phi_w' = -\kappa\, \phi_f'$ (3e)

on CD $\quad \phi_w' = 0$ (3f)

on DE $\quad \phi_w' = Bi_1 (1-\phi_w)$ (3g)

on EF $\quad \phi_w' = 0$ (3h)

on FO $\quad \phi_w' = -Bi_2\, \phi_w$ (3i)

where the prime (') denotes the derivative in the direction of the outward normal to the associated surface.

Conditions (3c), (3f) and (3h) arise from the symmetry of the fin array and stipulate that there is no heat flux across the fictitious boundaries BC, CD and EF, Fig. 1. Conditions (3d) and (3e) result from the assumption of perfect wall-to-fin contact which requires that the temperature and heat flux be continuous at the contact interface OC. The remaining conditions describe the convective exchange from the exposed surfaces FOAB and DE.

From the equations (1), (2) and (3) it can be deduced that the heat flow through the assembly may be parameterised by the Biot numbers, Bi_1 and Bi_2, the ratio of the thermal conductivities κ, and the aspect ratios L, T and W.

METHOD OF SOLUTION

BIE Formulation

As there is an extensive range of published literature giving detailed descriptions of the BIE procedure for solving plane potential boundary-value problems, e.g. [14-18], only those features necessary to facilitate a concise explanation of the BIE solution of the fin assembly problem, described by equations (1), (2) and (3), are presented here.

The BIE method is based upon Green's Integral Formula [14] which, for any sufficiently smooth function ϕ which satisfies Laplace's equation within a plane domain Ω having a piecewise-smooth boundary $\partial\Omega$, may be expressed as,

$$\int_{\partial\Omega} \{\phi(q) \log' |p-q| - \phi'(q) \log |p-q|\} \, dq = \eta(p)\phi(p), \qquad (4)$$

where

i $p \in \Omega + \partial\Omega$ and $q \in \partial\Omega$,

ii dq denotes the differential increment of $\partial\Omega$ at q,

iii the prime (') denotes the derivative in the direction of the outward normal to $\partial\Omega$ at q,

iv if $p \in \Omega$ then $\eta = 2\pi$, but if $p \in \partial\Omega$ then η is the angle included between the tangents to $\partial\Omega$ on either side of p.

If either ϕ or ϕ' is prescribed at each point of $\partial\Omega$, then the solution of the integral equation,

$$\int_{\partial\Omega} \{\phi(q) \log' |\bar{q}-q| - \phi'(q) \log |\bar{q}-q|\} \, dq - \eta(\bar{q})\phi(\bar{q}) = 0$$

$$q, \bar{q} \in \partial\Omega \qquad (5)$$

determines the boundary distributions of both ϕ and ϕ'. The potential ϕ at any interior point can then be computed employing equation (4).

If on some section of $\partial\Omega$ the boundary conditions specify neither ϕ nor ϕ', but stipulate a linear relation of the form,

$$\alpha(q) \phi'(q) + \beta(q) \phi(q) + \gamma(q) = 0$$

where α, β and γ are prescribed functions, then in equation (5) ϕ' is appropriately replaced and ϕ is treated as the relevant unknown. Thus,

Green's Boundary Formula, equation (5), enables well-posed Laplacian boundary-value problems to be reformulated as integral equations in which the unknowns are the boundary values of ϕ and ϕ' complementary to those prescribed by the boundary conditions.

In practice the integral equation (5) can rarely be solved analytically and therefore various numerical techniques have been proposed for effecting its solution, e.g. [14-18]. In this study the classical or piecewise-constant BIE discretisation is described. The techniques for implementing higher order BIE approximations can be found, for example, in Fairweather et al [17].

In the classical BIE method the boundary $\partial \Omega$ is subdivided into N smooth segments, $\partial \Omega_j$, $j = 1, \ldots, N$, on which ϕ and ϕ' are approximated by piecewise-constant functions ϕ_j and ϕ'_j. Correspondingly, Green's Integral Formula, equation (4), becomes,

$$\sum_{j=1}^{N} \{\phi_j \int_{\partial \Omega_j} \log'|p-q|\, dq - \phi'_j \int_{\partial \Omega_j} \log|p-q|\, dq\} = \eta(p)\, \phi(p)$$

$$p \in \Omega + \partial\Omega, \quad q \in \partial\Omega. \qquad (6)$$

Applying equation (6) to the midpoint q_j of each segment generates the system of algebraic equations,

$$A*\Phi + B*\Phi' = 0 \qquad (7)$$

where

$$A^* = (A^*_{ij}),\quad A^*_{ij} = \int_{\partial\Omega_j} \log'|q_i-q|\, dq - \delta_{ij}\, \eta(q_i) \qquad (8)$$

$$B^* = (B^*_{ij}),\quad B^*_{ij} = -\int_{\partial\Omega_j} \log|q_i-q|\, dq \qquad (9)$$

$$\phi = (\phi_1, \ldots, \phi_N)^T \qquad (10)$$

$$\phi' = (\phi_1', \ldots, \phi_N')^T \qquad (11)$$

and δ_{ij} denotes the Kronecker-delta function.

Introducing the boundary conditions into equation (7) yields a system of N linear-algebraic equations in which the unknowns are the boundary values of the ϕ_j and ϕ_j' complementary to those prescribed. The solution of this algebraic representation determines the boundary distributions of both ϕ_j and ϕ_j'. The solution at any point $p \in \Omega + \partial\Omega$ can then be computed by a relatively simple quadrature, equation (6).

BIE Application

The application of Green's Integral Formula, equation (4), to the problem described by equations (1), (2) and (3) gives rise to a pair of coupled integral equations,

$$\int_{\partial\Omega_f} \{\phi_f(q) \log'|p-q| - \phi_f'(q) \log|p-q|\} dq = \eta(p)\phi_f(p),$$

$$p \in \Omega_f + \partial\Omega_f, \quad q \in \partial\Omega_f \qquad (12)$$

and

$$\int_{\partial\Omega_w} \{\phi_w(s) \log'|r-s| - \phi_w'(s) \log|r-s|\} ds = \eta(r)\phi_w(r),$$

$$r \in \Omega_w + \partial\Omega_w, \quad s \in \partial\Omega_w \qquad (13)$$

where $\partial\Omega_f$ and $\partial\Omega_w$ denote the contours OABCO and OCDEFO, respectively, and Ω_f and Ω_w denote the regions bounded by OABCO and OCDEFO, respectively. The coupling results from the boundary conditions imposed at the wall-to-fin interface OC.

The contours $\partial\Omega_f$ and $\partial\Omega_w$ are subdivided into a total of N rectilinear segments and the temperature ϕ and its normal derivative ϕ' are approximated by piecewise-constant functions on each of these segments. Then, equation (12) is collocated at the midpoint of each segment of $\partial\Omega_f$ and equation (13) is collocated at the midpoint of each segment of $\partial\Omega_w$. This generates a system of N linear algebraic equations in the 2N unknown boundary values of ϕ and ϕ'. Subsequent inclusion of the boundary conditions (3) reduces the number of unknowns to N and gives rise to a system of linear algebraic equations of the form,

$$A \underline{x} = \underline{b} \qquad (14)$$

where the elements of the N x N matrix A and the vector $\underline{b}$ are formed by appropriate combinations of A^* and B^*, equations (9) and (10), and $\underline{x} = (x_1,\ldots,x_N)^T$ are the N unknown boundary values of ϕ and ϕ'.

The matrices A^* and B^* form the basic components of the BIE discretization process, yet the elements of these matrices depend only upon the configuration and dimensions of the domain boundary and the distribution of segments over this boundary. Introduction of the boundary conditions into the BIE solution procedure only involves simple algebraic manipulations of the matrices A^* and B^*. Consequently, the BIE formulation can easily handle quite general linear boundary conditions. Thus, in the context of fin heat transfer problems, the relaxation of the uniform heat transfer coefficient assumption can be achieved with the minimal additional programming effort because only the respective boundary conditions are affected. For example, if the variation of the heat transfer coefficient over the fin profile is prescribed then, on each segment of the side OA, the local heat transfer coefficient should be approximated by its nodal value and the appropriate combination of A^* and B^* be employed in the construction of the algebraic representation (14).

The system of equations (14) is considerably smaller than that generated by an equivalent finite-difference or finite-element representation because the BIE discretization occurs only on boundary of the domain within which the solution is required. Consequently, even for fine boundary discretizations, it is most appropriate to solve the system of equations (14) by a direct method such as Gaussian Elimination [19]. In fact, in previous applications of the BIE method, e.g. [15-18], solution of the corresponding algebraic equations have invariably been obtained by direct methods. Nevertheless, various iterative techniques are also suitable for effecting the solution to systems of simultaneous linear algebraic equations [19]. Therefore, in the context of the fin assembly problem, the performance of the method of Successive-Over-Relaxation [19] has been investigated. It was found that the iterative process fails to converge irrespective of the values of the system parameters, the size of the boundary discretization, and the magnitude of the relaxation parameter.

The solution of the equations (14) determines the boundary distribution of both the temperature and its normal derivative. The temperature at any point within the fin assembly can then be computed by simply integrating Green's Integral Formula around the boundary of the region in which that point lies, as indicated in equation (6). In particular, the temperature at points on the common interface OC can be obtained by integrating around either $\partial\Omega_f$ or $\partial\Omega_w$, thus providing a means for checking the accuracy of the BIE solution.

FIN ASSEMBLY HEAT TRANSFER RATE

The heat flow rate through the fin assembly is most conveniently expressed in the form of an augmentation factor, Aug, defined as the ratio of the heat transfer rate of the fin assembly to that of the unfinned wall operating under the same conditions. This augmentation factor can be evalauted at either of the exposed surfaces DE and FOAB and is given by,

$$\text{Aug} = \left[\frac{1}{Bi_1} + W + \frac{1}{Bi_2}\right] \int_{DE} \phi_w'(q)\, dq \qquad (15)$$

$$\equiv \left[\frac{1}{Bi_1} + W + \frac{1}{Bi_2}\right] \left\{ \int_{FO} \phi_w'(q)\, dq + \kappa \int_{OAB} \phi_f'(q)\, dq \right\} \qquad (16)$$

In the context of the BIE solutions these integrations can be performed numerically in a manner consistent with the overall discretization. However, as the BIE solution is only an approximation it need not give exactly the same values for the integrations (15) and (16), although, for the solution to be satisfactory, these should agree to within an acceptable tolerance. Therefore, in the subsequent BIE calculations Aug_1 and Aug_2 shall denote the values of the augmentation factor corresponding to the expressions (15) and (16), respectively.

With the finite-difference and finite-element methods the evaluation of the expressions (15) and (16) requires the determination of the temperature distribution throughout the domain $\Omega_f + \Omega_w$. In contrast, the BIE method yields all the necessary information for the computation of these quantities, namely the boundary distribution of ϕ', when the boundary integral representation of the problem described by equations (1), (2), and (3) is solved. With the BIE formulation the temperature distribution within the domain $\Omega_f + \Omega_w$ is simply generated from the boundary distributions of ϕ and ϕ', and need only be computed if desired.

BIE RESULTS

Solutions have been obtained for a wide range of the system parameters Bi_1, Bi_2, κ, L, T and W. For each particular problem solutions were computed employing 80, 160 and 320 boundary segments. The distribution of these segments accounted for the fact that in practice the fin length is considerably larger than the fin thickness, the fin-pitch and the wall thickness, and accordingly, the sides OA and BC were subdivided into four times as many segments as the sides AB, CO, OC, CD, DX, XE, EF and FO, Fig. 1. This particular boundary discretization was found to offer the most efficient use of the computational resources with respect to the accuracy of the corresponding solutions. In contrast to using 80, 160 and 320 boundary segments the equivalent finite-difference and finite-element implementations would require 192, 682 and 2562 nodes, respectively, distributed over the domain $(\Omega_f + \partial\Omega_f) + (\Omega_w + \partial\Omega_w)$.

In order to illustrate the performance of the BIE method the results for three particular problems are presented in Tables 1, 2 and 3. These Tables show the dimensionless temperature distribution ϕ at the lattice points of a uniform mesh and also the respective values of the augmentation factors Aug_1 and Aug_2. These results correspond to the problems,

1. $Bi_1 = 1.00$, $Bi_2 = 0.01$, $\kappa = 1.0$, $L = 10.0$, $T = 0.5$ and $W = 5.0$

2. $Bi_1 = 2.25$, $Bi_2 = 0.75$, $\kappa = 1.0$, $L = 10.0$, $T = 0.5$ and $W = 5.0$

3. $Bi_1 = 2.25$, $Bi_2 = 0.75$, $\kappa = 10.0$, $L = 10.0$, $T = 0.5$ and $W = 5.0$

Problem 1 represents the performance of a stainless-steel finned heat exchanger with forced convection of water on the plain side and free convection of air on the fin side. The system parameters for problems 2 and 3 lie in the range $Bi_1 > 2.0$ and $Bi_2 > 0.5$ for which Stones [11] has indicated that neither the finite-difference method nor the finite-element method produce acceptable solutions.

The results presented in Tables 1, 2 and 3 illustrate various features of the BIE solution technique and are characteristic of those observed for other values of the system parameters. In particular, in all cases considered the temperature distribution ϕ and the augmentation factors, Aug_1 and Aug_2, display a convergent behaviour as the boundary discretization is refined. In addition, the accuracy with which the energy conservation criteria, i.e. Aug_1 be identically equal to Aug_2, is satisfied improves as the number of boundary segments is increased. Closer inspection of the results in Tables 1, 2 and 3 reveals a fundamental weakness of the classical BIE method in that at each of the corners of the domain $\Omega_f + \Omega_w$ the temperature is either slightly depressed or slightly elevated in comparison to that at neighbouring points. This inaccuracy is a consequence of the relatively crude assumed boundary variations of ϕ and ϕ', and should be improved by the use of more sophisticated approximations [16,17,18]. Solutions to problem 1 have therefore been computed employing piecewise-linear [18] and piecewise-quadratic [16,17] BIE implementations and these solutions are displayed in Tables 4 and 5, respectively. It is clearly evident from these solutions that the higher order approximations resolve the difficulties encountered by the classical BIE method at the domain corners and at the same time facilitate an improvement in the accuracy with which the temperature continuity boundary condition (3d) is satisfied. However, the additional sophistications inherent in the higher order BIE methods necessitate considerably more programming effort than the classical method, although, for a given number of boundary nodes the computational requirements of all three BIE implementations are approximately the same.

The results in Tables 1-5 show that, as the boundary discretization is uniformly refined from 80 to 320 segments, the BIE solutions display a convergent behaviour but do not actually achieve the limiting solution. The primary cause for this slow convergence is the presence of a boundary singularity at the re-entrant corner O which, in fact, will have similar effects on solutions computed by the standard finite-difference and finite-element implementations. As further refinement of the boundary discretization is impractical, the limiting values of the augmentation factors, Aug_1 and Aug_2, have been obtained by extrapolation, using Richardson's formula [19],

$$E(N) \alpha H^{\alpha}(N)$$

where $E(N)$ is the error of the solution given by the N segment discretization, $H(N)$ is an associated segment length and α is the order of the extrapolation. The excellent agreement between the extrapolated values of Aug_1 and Aug_2, Tables 1-5, emphasises the compatability of Richardson's extrapolation method with the BIE solutions.

BIE TREATMENT OF BOUNDARY SINGULARITIES

Symm [15] and Papamichael and Symm [20] have shown for problems with more simplified boundary conditions than those pertaining to the fin assembly model, that by incorporating the analytical behaviour of the singularity into the classical BIE method it is possible to substantially reduce the effect of the singularity and thereby to obtain solutions which converge appreciably more rapidly.

Following the procedure described by Motz [21] it can be shown that in the vicinity of the re-entrant corner O, Fig. 1,

$$\phi_f(r,\xi) = \alpha(1 - \frac{Bi_2}{\kappa} r\cos\xi + \frac{Bi_2}{\kappa} r\sin\xi - \frac{1}{2}(\frac{Bi_2}{\kappa})^2 r^2\sin 2\xi)$$

$$+ \beta(r^2\cos 2\xi) + \gamma(r^{\lambda_o}\cos\lambda_o\xi + \frac{1}{\lambda_1}\frac{Bi_2}{\kappa}r^{\lambda_1}\sin\lambda_1\xi)$$

$$+ \gamma_o r^{\lambda_1}\cos\lambda_1\xi) + \delta(r^{\lambda_2}\cos\lambda_2\xi) + \ldots \tag{17}$$

and

$$\phi_w(r,\eta) = \alpha(1 - \frac{Bi_2}{\kappa} r\cos\eta + Bi_2 r\sin\eta - \frac{1}{2}\frac{Bi_2^2}{\kappa} r^2\sin 2\eta)$$

$$+ \beta(-r^2\cos 2\eta) + \gamma(\gamma_1 r^{\lambda_o}\cos\lambda_o\eta + \frac{\gamma_1}{\lambda_1}Bi_2 r^{\lambda_1}\sin\lambda_1\eta)$$

$$+ \gamma_2 r^{\lambda_1}\cos\lambda_1\eta) + \delta(\delta_o r^{\lambda_2}\cos\lambda_2\eta) + \ldots \tag{18}$$

where (r,ξ) and (r,η) are the polar coordinate systems as indicated in Fig. 1, α, β, γ and δ are unknown constants,

$$\gamma_o = \cos\lambda_o \frac{\pi}{2} / \cos\lambda_o \pi$$

$$\gamma_1 = (\Gamma_o \sin\lambda_o \pi + \Gamma_1 \cos\lambda_o \pi)/\Gamma_2$$

$$\gamma_2 = (\Gamma_o \kappa \cos\lambda_o \frac{\pi}{2} + \Gamma_1 \sin\lambda_o \frac{\pi}{2})/\Gamma_2$$

$$\delta_o = \cos\lambda_2 \frac{\pi}{2} / \cos\lambda_2 \pi$$

$$\Gamma_o = \frac{Bi_2}{\kappa} \cos\lambda_o \frac{\pi}{2} - Bi_2 \kappa \sin\lambda_o \frac{\pi}{2}$$

$$\Gamma_1 = Bi_2 \sin\lambda_o \frac{\pi}{2} + Bi_2 \cos\lambda_o \frac{\pi}{2}$$

$$\Gamma_2 = \lambda_1(\sin\lambda_o \frac{\pi}{2} \sin\lambda_o \pi - \kappa\cos\lambda_o \frac{\pi}{2} \cos\lambda_o \pi)$$

$$\lambda_o = \frac{2}{\pi} \cos^{-1}\left(\left(\frac{1}{2(1+\kappa)}\right)^{\frac{1}{2}}\right)$$

$$\lambda_1 = 1 + \lambda_o$$

and

$$\lambda_2 = 2 - \lambda_o .$$

From these expressions it can be deduced that, irrespective of the values of the system parameters, all derivatives of ϕ_f and ϕ_w become unbounded as the point O is approached.

The singular BIE method [15] involves the inclusion of the first 4 terms of each of the expansions (17) and (18) into the classical BIE method. The performance of this singular BIE technique has been investigated and the results indicate that only for problems for which the aspect ratio L is less than unity does the inclusion of the analytical form of the singularity improve the rate of convergence. This can be attributed to the fact that for larger (more practical) values of the parameter L the approximations associated with the inclusion of the singular terms into the BIE solution procedure become inapplicable.

DISCUSSION AND CONCLUSIONS

The advanced solution capabilities offered by the BIE formulation have been demonstrated by the application of the piecewise-constant, piecewise-linear, piecewise-quadratic and singular BIE methods to a fin assembly heat transfer problem. The BIE methods have several computational advantages over comparable finite-difference and finite-element schemes. In particular, the BIE discretization occurs only on the boundary of the relevant domain and therefore generates a considerably smaller system of equations than equivalent finite-difference or finite-element representations.

The additional sophistications of the higher order BIE implementations, while requiring considerably more programming time than the classical BIE method, facilitate improved accuracy. However, the singular BIE method described by Symm [15] is inappropriate for the fin assembly problem, because the approximations associated with this method become invalid for realistic values of the system parameters.

The suitability of the BIE method for the solution of fin assembly heat transfer problems is confirmed by the consistency of the results predicted by the 3 different BIE implementations, (Tables 1, 4 and 5) and the fact that the extrapolated values of Aug_1 and Aug_2 agree to at least 3 significant figures. Furthermore, the BIE method gives accurate solutions even for problems for which the finite-difference and finite-element methods fail to provide acceptable results, Tables 2 and 3.

In this study the application of the BIE method has been restricted to plane problems involving rectangular fins and constant heat transfer coefficients. However, it must be emphasised that the BIE formulation can easily handle curved fin profiles and non-uniform variations of surface heat transfer. The extension of the work presented in this study to heat transfer problems involving annular finned tubing should be accomplished without difficulty by employing an axisymmetric BIE formulation [22].

NOMENCLATURE

Aug	augmentation factor
Bi_1	$= h_1 P/k_w$, Biot number
Bi_2	$= h_2 P/k_w$, Biot number
h_1, h_2	heat transfer coefficients, $W/m^2 K$
k_f, k_w	thermal conductivities, W/mK
l	fin length, m
L	$= l/P$, aspect ratio
P	half fin pitch, m
t	half fin thickness, m
T	$= t/P$, aspect ratio
w	wall thickness, m
W	$= w/P$, aspect ratio
κ	$= k_f/k_w$
θ	temperature distribution, K
θ_1, θ_2	fluid temperatures, K
ϕ	$= (\theta - \theta_2)/(\theta_1 - \theta_2)$, dimensionless temperature distribution

Subscripts

1	plain side
2	fin side
f	fin
w	wall

REFERENCES

1. K.A. Gardner, "Efficiency of Extended Surface", Transactions of the ASME, Vol. 67, pp. 621-631, 1945.

2. D.Q. Kern and A.D. Kraus, Extended Surface Heat Transfer, McGraw-Hill, New York, 1972.

3. I. Mikk, "Convective Fin of Minimum Mass", International Journal of Heat and Mass Transfer, Vol. 23, pp. 707-711, 1980.

4. R.K. Irey, "Errors in the One-Dimensional Fin Solution", Journal of Heat Transfer, Vol. 90, pp. 175-176, 1968.

5. M. Levitsky, "The Criterion for the Validity of the Fin Approxmation", International Journal of Heat and Mass Transfer, Vol. 15, pp. 1960-1963, 1972.

6. W. Lau and C.W. Tan, "Errors in One-Dimensional Heat Transfer Analysis in Straight and Annular Fins", Journal of Heat Transfer, Vol. 95, pp. 549-551, 1973.

7. E.M. Sparrow and L. Lee, "Effects of Fin Base-Temperature Depression in a Multifin Array", Journal of Heat Transfer, Vol. 97, pp. 463-465, 1975.

8. N.V. Suryanarayana, "Two-Dimensional Effects on Heat Transfer Rates From an Array of Straight Fins", Journal of Heat Transfer, Vol. 99, pp. 129-132, 1977.

9. P.J. Heggs and P.R. Stones, "The Effects of Dimensions on the Heat Flow Rate Through Extended Surfaces", Journal of Heat Transfer, Vol. 102, pp. 180-182, 1980.

10. P.J. Heggs and P.R. Stones, "Improved Design Methods for Finned Tube Heat Exchangers", Transactions of the Institution of Chemical Engineers, Vol. 58, pp. 147-154, 1980.

11. P.R. Stones, Ph.D. Thesis, Leeds University, England, 1980.

12. G.D. Smith, Numerical Solution of Partial Differential Equations, Oxford University Press, 1974.

13. O.C. Zienkiewicz, The Finite Element Method in Engineering, McGraw-Hill, London, 1971.

14. M.A. Jaswon and G.T. Symm, Integral Equation Methods in Potential Theory and Electrostatics, Academic Press, London, 1977.

15. G.T. Symm, "Treatment of Singularities in the Solution of Laplace's Equation by an Integral Equation Method", National Physical Laboratory, Report NAC31, 1973.

16. Y.S. Wu, F.J. Rizzo, D.J. Shippy and J.A. Wagner, "An Advanced Boundary Integral Equation Method for Two-Dimensional Electromagnetic Field Problems", Electric Machines and Electromechanics, Vol. 1, pp. 301-313, 1977.

17. G. Fairweather, F.J. Rizzo, D.J. Shippy and Y.S. Wu, "On the Numerical Solution of Two-Dimensional Potential Problems by an Improved Boundary Integral Equation Method", Journal of Computational Physics, Vol. 31, pp. 96-112, 1979.

18. R.F. Harrington, K. Pontoppidan, P. Abrahamsen and N.C. Albertsen, "Computation of Laplacian Potentials by an Equivalent Source Method", Proceedings of the IEEE, Vol. 116, pp. 1715-1719, 1969.

19. A. Ralston, A First Course in Numerical Analysis, McGraw-Hill, New York, 1965.

20. N. Papamichael and G.T. Symm, "Numerical Techniques for Two-Dimensional Laplacian Problems", Computer Methods in Applied Mechanics and Engineering, Vol. 6, pp. 175-194, 1975.

21. H. Motz, "The Treatment of Singularities of Partial Differential Equations by Relaxation Methods", Quarterly of Applied Mathematics, Vol. 4, pp. 371-377, 1946.

22. F.J. Rizzo and D.J. Shippy, "A Boundary Integral Equation Approach to Potential and Elasticity Problems for Axisymmetric Bodies with Arbitrary Boundary Conditions", Mechanics Research Communications, Vol. 6, pp. 99-103, 1979.

Table 1: Classical BIE Method Solutions for the Problem 1

D										B
0.9583 0.9542 0.9526	0.8625 0.8576 0.8560	0.8161 0.8099 0.8079	0.7693 0.7617 0.7592	0.7282 0.7112 0.7040	0.5745 0.5574 0.5513	0.4662 0.4488 0.4425	0.3930 0.3754 0.3690	0.3496 0.3315 0.3250	0.3323 0.3136 0.3068	
0.9490 0.9505 0.9514	0.9084 0.9051 0.9040	0.8625 0.8576 0.8560	0.8161 0.8099 0.8079	0.7692 0.7617 0.7592	0.7192 0.7060 0.7011	0.5745 0.5573 0.5512	0.4663 0.4488 0.4425	0.3930 0.3754 0.3690	0.3496 0.3315 0.3249	0.3324 0.3136 0.3068
0.9508 0.9516 0.9518	0.9084 0.9051 0.9040	0.8625 0.8576 0.8560	0.8162 0.8099 0.8079	0.7694 0.7618 0.7593	0.7186 0.7060 0.7014	0.5743 0.5572 0.5510	0.4661 0.4486 0.4423	0.3929 0.3752 0.3689	0.3495 0.3314 0.3248	0.3322 0.3135 0.3067
0.9517 0.9520 0.9519	0.9083 0.9051 0.9040	0.8625 0.8576 0.8560	0.8162 0.8099 0.8079	0.7697 0.7620 0.7595	0.7187 0.7065 0.7023	0.5740 0.5569 0.5508	0.4659 0.4484 0.4421	0.3927 0.3751 0.3687	0.3493 0.3312 0.3247	0.3321 0.3133 0.3065
0.9522 0.9522 0.9520	0.9083 0.9051 0.9040	0.8625 0.8576 0.8560	0.8162 0.8099 0.8079	0.7701 0.7622 0.7597	0.7195 0.7076 0.7039	0.5736 0.5565 0.5504	0.4656 0.4481 0.4418	0.3925 0.3748 0.3684	0.3491 0.3310 0.3245	0.3319 0.3131 0.3063
0.9524 0.9523 0.9520	0.9083 0.9051 0.9040	0.8625 0.8576 0.8560	0.8162 0.8099 0.8079	0.7704 0.7624 0.7599	0.7287 0.7144 0.7095	0.5730 0.5560 0.5499	0.4651 0.4477 0.4414	0.3920 0.3745 0.3681	0.3487 0.3307 0.3242	0.3314 0.3128 0.3060
0.9523 0.9522 0.9520	0.9083 0.9051 0.9040	0.8625 0.8576 0.8560	0.8162 0.8099 0.8079	0.7707 0.7626 0.7601	0.7325 0.7213 0.7180					A
0.9518 0.9520 0.9519	0.9084 0.9051 0.9040	0.8625 0.8576 0.8560	0.8163 0.8099 0.8079	0.7709 0.7628 0.7603	0.7367 0.7255 0.7222					
0.9508 0.9516 0.9518	0.9084 0.9051 0.9040	0.8625 0.8576 0.8560	0.8163 0.8099 0.8079	0.7711 0.7630 0.7604	0.7396 0.7281 0.7247					
0.9490 0.9506 0.9514	0.9085 0.9051 0.9040	0.8625 0.8576 0.8560	0.8163 0.8100 0.8079	0.7713 0.7631 0.7605	0.7419 0.7298 0.7261					
0.9583 0.9542 0.9526	0.9086 0.9052 0.9040	0.8625 0.8576 0.8560	0.8162 0.8099 0.8079	0.7709 0.7630 0.7605	0.7371 0.7288 0.7262					
E					F					

	N=80	N=160	N=320	limit
Aug_1	5.2303	5.1556	5.1286	5.1133
Aug_2	5.4241	5.2246	5.1530	5.1128
Aug_1/Aug_2	0.9643	0.9868	0.9953	1.0001

$$N = \begin{cases} 80 \\ 160 \\ 320 \end{cases}$$

Table 2 : Classical BIE Method Solutions for the Problem 2

D				C				B
0.9534 0.9400 0.9347	0.7831 0.7794 0.7782	0.6268 0.6246 0.6239	0.4696 0.4694 0.4695	0.3128 0.3142 0.3149	0.1893 0.1736 0.1659	0.0196 0.0175 0.0168	0.0027 0.0020 0.0018	0.0009 0.0004 0.0003	0.0014 0.0005 0.0002
0.9209 0.9275 0.9304	0.7827 0.7793 0.7781	0.6267 0.6245 0.6239	0.4697 0.4694 0.4695	0.3130 0.3143 0.3149	0.1702 0.1622 0.1594	0.0197 0.0174 0.0167	0.0027 0.0020 0.0018	0.0009 0.0004 0.0003	0.0012 0.0005 0.0002
0.9262 0.9304 0.9315	0.7825 0.7793 0.7781	0.6267 0.6245 0.6239	0.4698 0.4694 0.4695	0.3134 0.3144 0.3150	0.1662 0.1600 0.1580	0.0192 0.0171 0.0164	0.0026 0.0019 0.0017	0.0008 0.0004 0.0002	0.0012 0.0004 0.0002
0.9289 0.9316 0.9318	0.7823 0.7792 0.7781	0.6267 0.6245 0.6239	0.4698 0.4694 0.4695	0.3137 0.3145 0.3151	0.1629 0.1578 0.1564	0.0186 0.0165 0.0158	0.0025 0.0019 0.0017	0.0008 0.0004 0.0002	0.0011 0.0004 0.0002
0.9302 0.9320 0.9319	0.7822 0.7792 0.7781	0.6267 0.6245 0.6239	0.4698 0.4695 0.4695	0.3140 0.3147 0.3152	0.1597 0.1549 0.1545	0.0177 0.0157 0.0151	0.0024 0.0018 0.0016	0.0008 0.0004 0.0002	0.0011 0.0004 0.0002
0.9307 0.9322 0.9319	0.7822 0.7792 0.7781	0.6267 0.6245 0.6239	0.4698 0.4695 0.4695	0.3142 0.3148 0.3153	0.1711 0.1609 0.1576	0.0164 0.0147 0.0141	0.0022 0.0017 0.0015	0.0007 0.0004 0.0002	0.0013 0.0004 0.0002
0.9302 0.9320 0.9319	0.7822 0.7792 0.7781	0.6267 0.6245 0.6239	0.4698 0.4695 0.4695	0.3144 0.3149 0.3154	0.1694 0.1646 0.1642				
0.9289 0.9316 0.9318	0.7823 0.7792 0.7781	0.6267 0.6245 0.6239	0.4698 0.4695 0.4696	0.3146 0.3150 0.3155	0.1742 0.1684 0.1675				
0.9262 0.9304 0.9315	0.7825 0.7793 0.7781	0.6267 0.6245 0.6239	0.4698 0.4695 0.4696	0.3146 0.3151 0.3156	0.1791 0.1715 0.1696				
0.9209 0.9275 0.9304	0.7827 0.7793 0.7781	0.6267 0.6245 0.6239	0.4698 0.4695 0.4696	0.3146 0.3151 0.3156	0.1855 0.1754 0.1716				
0.9534 0.9400 0.9347	0.7831 0.7794 0.7782	0.6268 0.6246 0.6239	0.4697 0.4696 0.4696	0.3140 0.3150 0.3156	0.1574 0.1653 0.1686				
E				F				A	

$N = \begin{array}{c} 80 \\ 160 \\ 320 \end{array}$

	N=80	N=160	N=320	limit
Aug_1	1.1344	1.0726	1.0528	1.0435
Aug_2	1.1608	1.0820	1.0561	1.0434
Aug_1 / Aug_2	0.9772	0.9913	0.9969	1.0001

Table 3 : Classical BIE Method Solutions for the Problem 3

Corner labels: D (top-left), C (top, between col 5-6), B (top-right), A (right), E (bottom-left), F (bottom-right).

0.6763	0.6009	0.5305	0.4598	0.3887	0.3279	0.1541	0.0744	0.0378	0.0221	0.0176
0.6790	0.6056	0.5340	0.4624	0.3901	0.3176	0.1476	0.0697	0.0340	0.0184	0.0134
0.6798	0.6073	0.5355	0.4635	0.3910	0.3124	0.1454	0.0680	0.0326	0.0171	0.0119
0.6623	0.6007	0.5305	0.4598	0.3885	0.3159	0.1542	0.0745	0.0378	0.0221	0.0175
0.6737	0.6055	0.5340	0.4624	0.3901	0.3105	0.1475	0.0697	0.0339	0.0184	0.0134
0.6781	0.6073	0.5355	0.4635	0.3911	0.3085	0.1452	0.0680	0.0326	0.0171	0.0119
0.6653	0.6006	0.5305	0.4598	0.3888	0.3146	0.1538	0.0743	0.0377	0.0220	0.0174
0.6754	0.6055	0.5340	0.4624	0.3903	0.3102	0.1472	0.0695	0.0339	0.0184	0.0134
0.6787	0.6073	0.5355	0.4636	0.3912	0.3086	0.1449	0.0678	0.0325	0.0171	0.0119
0.6670	0.6006	0.5305	0.4598	0.3892	0.3142	0.1532	0.0740	0.0375	0.0220	0.0174
0.6761	0.6055	0.5340	0.4624	0.3905	0.3104	0.1466	0.0692	0.0337	0.0183	0.0133
0.6790	0.6073	0.5355	0.4636	0.3913	0.3093	0.1444	0.0676	0.0324	0.0170	0.0118
0.6679	0.6005	0.5305	0.4599	0.3896	0.3146	0.1525	0.0736	0.0374	0.0219	0.0173
0.6765	0.6055	0.5340	0.4624	0.3907	0.3111	0.1459	0.0689	0.0336	0.0182	0.0133
0.6790	0.6073	0.5355	0.4636	0.3915	0.3106	0.1436	0.0672	0.0322	0.0169	0.0118
0.6682	0.6005	0.5305	0.4599	0.3899	0.3258	0.1513	0.0731	0.0371	0.0217	0.0172
0.6766	0.6055	0.5340	0.4624	0.3910	0.3190	0.1449	0.0684	0.0333	0.0181	0.0132
0.6791	0.6073	0.5355	0.4636	0.3918	0.3169	0.1427	0.0668	0.0320	0.0168	0.0117
0.6679	0.6005	0.5305	0.4599	0.3903	0.3300					
0.6765	0.6055	0.5340	0.4624	0.3912	0.3270					
0.6790	0.6073	0.5355	0.4636	0.3920	0.3268					
0.6670	0.6006	0.5305	0.4600	0.3906	0.3352					
0.6761	0.6055	0.5340	0.4624	0.3915	0.3319					
0.6790	0.6073	0.5355	0.4636	0.3922	0.3317					
0.6653	0.6006	0.5305	0.4600	0.3908	0.3390					
0.6754	0.6055	0.5340	0.4624	0.3916	0.3351					
0.6787	0.6073	0.5355	0.4636	0.3924	0.3346					
0.6624	0.6007	0.5305	0.4600	0.3910	0.3424					
0.6737	0.6055	0.5340	0.4624	0.3917	0.3376					
0.6781	0.6073	0.5355	0.4636	0.3925	0.3365					
0.6763	0.6009	0.5306	0.4599	0.3905	0.3336					
0.6790	0.6056	0.5340	0.4624	0.3917	0.3351					
0.6798	0.6073	0.5355	0.4636	0.3925	0.3362					

	$N=80$	$N=160$	$N=320$	limit
Aug_1	1.7158	1.6653	1.6478	1.6385
Aug_2	1.7709	1.6846	1.6547	1.6388
Aug_1/Aug_2	0.9689	0.9885	0.9958	0.9998

$N = 80$
160
320

Table 4 : Linear BIE Method Solutions for the Problem 1

D				C					B	
0.9512 0.9516 0.9517	0.9025 0.9032 0.9034	0.8538 0.8548 0.8551	0.8053 0.8064 0.8068	0.7565 0.7576 0.7579	0.6965 0.6976 0.6981	0.5437 0.5461 0.5472	0.4337 0.4368 0.4382	0.3592 0.3629 0.3645	0.3142 0.3184 0.3202	0.2951 0.2999 0.3018
0.9512 0.9516 0.9517	0.9025 0.9032 0.9034	0.8538 0.8548 0.8551	0.8053 0.8064 0.8068	0.7562 0.7575 0.7579	0.6963 0.6977 0.6982	0.5438 0.5461 0.5472	0.4337 0.4368 0.4381	0.3592 0.3629 0.3644	0.3142 0.3184 0.3202	0.2952 0.2999 0.3018
0.9512 0.9516 0.9517	0.9025 0.9032 0.9034	0.8538 0.8548 0.8551	0.8053 0.8064 0.8068	0.7563 0.7576 0.7580	0.6969 0.6983 0.6988	0.5436 0.5459 0.5470	0.4336 0.4367 0.4380	0.3591 0.3628 0.3643	0.3141 0.3183 0.3201	0.2951 0.2998 0.3017
0.9512 0.9516 0.9517	0.9025 0.9032 0.9034	0.8538 0.8548 0.8551	0.8053 0.8064 0.8068	0.7564 0.7577 0.7582	0.6981 0.6995 0.6999	0.5433 0.5457 0.5467	0.4334 0.4364 0.4378	0.3589 0.3626 0.3642	0.3140 0.3182 0.3199	0.2950 0.2996 0.3015
0.9512 0.9516 0.9517	0.9025 0.9032 0.9034	0.8538 0.8548 0.8551	0.8053 0.8064 0.8068	0.7566 0.7579 0.7584	0.7001 0.7015 0.7019	0.5429 0.5453 0.5463	0.4331 0.4361 0.4375	0.3587 0.3623 0.3639	0.3138 0.3180 0.3197	0.2947 0.2994 0.3013
0.9512 0.9516 0.9517	0.9025 0.9032 0.9034	0.8538 0.8548 0.8551	0.8052 0.8064 0.8068	0.7568 0.7581 0.7586	0.7040 0.7057 0.7064	0.5423 0.5448 0.5458	0.4326 0.4357 0.4371	0.3583 0.3620 0.3636	0.3134 0.3177 0.3194	0.2944 0.2991 0.3010 A
0.9512 0.9516 0.9517	0.9025 0.9032 0.9034	0.8538 0.8548 0.8551	0.8052 0.8064 0.8068	0.7570 0.7583 0.7588	0.7150 0.7163 0.7166					
0.9512 0.9516 0.9517	0.9024 0.9032 0.9034	0.8538 0.8548 0.8551	0.8052 0.8064 0.8068	0.7572 0.7585 0.7589	0.7191 0.7204 0.7207					
0.9512 0.9516 0.9517	0.9024 0.9032 0.9034	0.8538 0.8548 0.8551	0.8052 0.8064 0.8068	0.7573 0.7586 0.7591	0.7215 0.7229 0.7232					
0.9511 0.9516 0.9517	0.9024 0.9032 0.9034	0.8538 0.8548 0.8551	0.8052 0.8064 0.8068	0.7573 0.7587 0.7592	0.7226 0.7243 0.7246					
E 0.9511 0.9515 0.9517	0.9024 0.9032 0.9034	0.8538 0.8548 0.8551	0.8051 0.8064 0.8068	0.7569 0.7587 0.7592	0.7220 0.7242 0.7249 F					

	N=80	N=160	N=320	limit
Aug_1	5.1763	5.1344	5.1208	5.1142
Aug_2	5.0517	5.0881	5.1037	5.1151
Aug_1/Aug_2	1.0247	1.0091	1.0034	0.9998

$$N = \begin{array}{|c|} \hline 80 \\ 160 \\ 320 \\ \hline \end{array}$$

Table 5: Quadratic BIE Method Solutions for the Problem 1

D 0.9518	0.9036	0.8554	0.8071	0.7581	0.6977	0.5447	0.4351	0.3610	0.3164	0.2977
0.9517	0.9035	0.8552	0.8069	0.7580	0.6979	0.5466	0.4374	0.3637	0.3194	0.3009
0.9518	0.9035	0.8552	0.8070	0.7581 C	0.6982	0.5474	0.4384	0.3648	0.3206 B	0.3022
0.9518	0.9036	0.8554	0.8071	0.7584	0.6977	0.5447	0.4350	0.3610	0.3164	0.2977
0.9517	0.9035	0.8552	0.8069	0.7581	0.6981	0.5465	0.4374	0.3637	0.3193	0.3009
0.9518	0.9035	0.8552	0.8070	0.7581	0.6984	0.5474	0.4384	0.3648	0.3205	0.3021
0.9518	0.9036	0.8554	0.8071	0.7584	0.6986	0.5445	0.4349	0.3609	0.3163	0.2976
0.9517	0.9035	0.8552	0.8069	0.7582	0.6987	0.5464	0.4373	0.3635	0.3192	0.3008
0.9518	0.9035	0.8552	0.8070	0.7582	0.6990	0.5472	0.4383	0.3646	0.3204	0.3021
0.9518	0.9036	0.8554	0.8071	0.7585	0.7000	0.5443	0.4347	0.3607	0.3161	0.2975
0.9517	0.9035	0.8552	0.8069	0.7583	0.6999	0.5461	0.4371	0.3634	0.3191	0.3006
0.9518	0.9035	0.8552	0.8070	0.7584	0.7001	0.5469	0.4380	0.3645	0.3203	0.3019
0.9518	0.9036	0.8554	0.8071	0.7586	0.7014	0.5439	0.4344	0.3604	0.3159	0.2972
0.9517	0.9035	0.8552	0.8069	0.7585	0.7018	0.5457	0.4367	0.3631	0.3188	0.3004
0.9518	0.9035	0.8552	0.8070	0.7585	0.7020	0.5465	0.4377	0.3642	0.3200	0.3017
0.9518	0.9036	0.8554	0.8071	0.7588	0.7058	0.5434	0.4340	0.3601	0.3156	0.2970
0.9517	0.9035	0.8552	0.8069	0.7587	0.7062	0.5452	0.4364	0.3628	0.3186	0.3001
0.9518	0.9035	0.8552	0.8070	0.7587	0.7065	0.5461	0.4373	0.3639	0.3198 A	0.3014
0.9518	0.9036	0.8554	0.8071	0.7590	0.7175					
0.9517	0.9035	0.8552	0.8070	0.7589	0.7170					
0.9518	0.9035	0.8552	0.8070	0.7589	0.7169					
0.9518	0.9036	0.8554	0.8071	0.7591	0.7215					
0.9517	0.9035	0.8552	0.8070	0.7590	0.7211					
0.9518	0.9035	0.8552	0.8070	0.7591	0.7210					
0.9518	0.9036	0.8554	0.8071	0.7592	0.7240					
0.9517	0.9035	0.8552	0.8070	0.7592	0.7236					
0.9518	0.9035	0.8552	0.8070	0.7593	0.7234					
0.9518	0.9036	0.8554	0.8071	0.7592	0.7253					
0.9517	0.9035	0.8552	0.8070	0.7593	0.7249					
0.9518	0.9035	0.8553	0.8070	0.7594	0.7248					
0.9518	0.9036	0.8554	0.8071	0.7595	0.7254					
0.9517	0.9035	0.8552	0.8070	0.7594	0.7251					
0.9518 E	0.9035	0.8553	0.8070	0.7594	0.7252 F					

	N=80	N=160	N=320	limit
Aug_1	5.1067	5.1160	5.1143	5.1146
Aug_2	5.0693	5.0957	5.1068	5.1148
Aug_1/Aug_2	1.0074	1.0039	1.0015	1.0000

N = 80
160
320

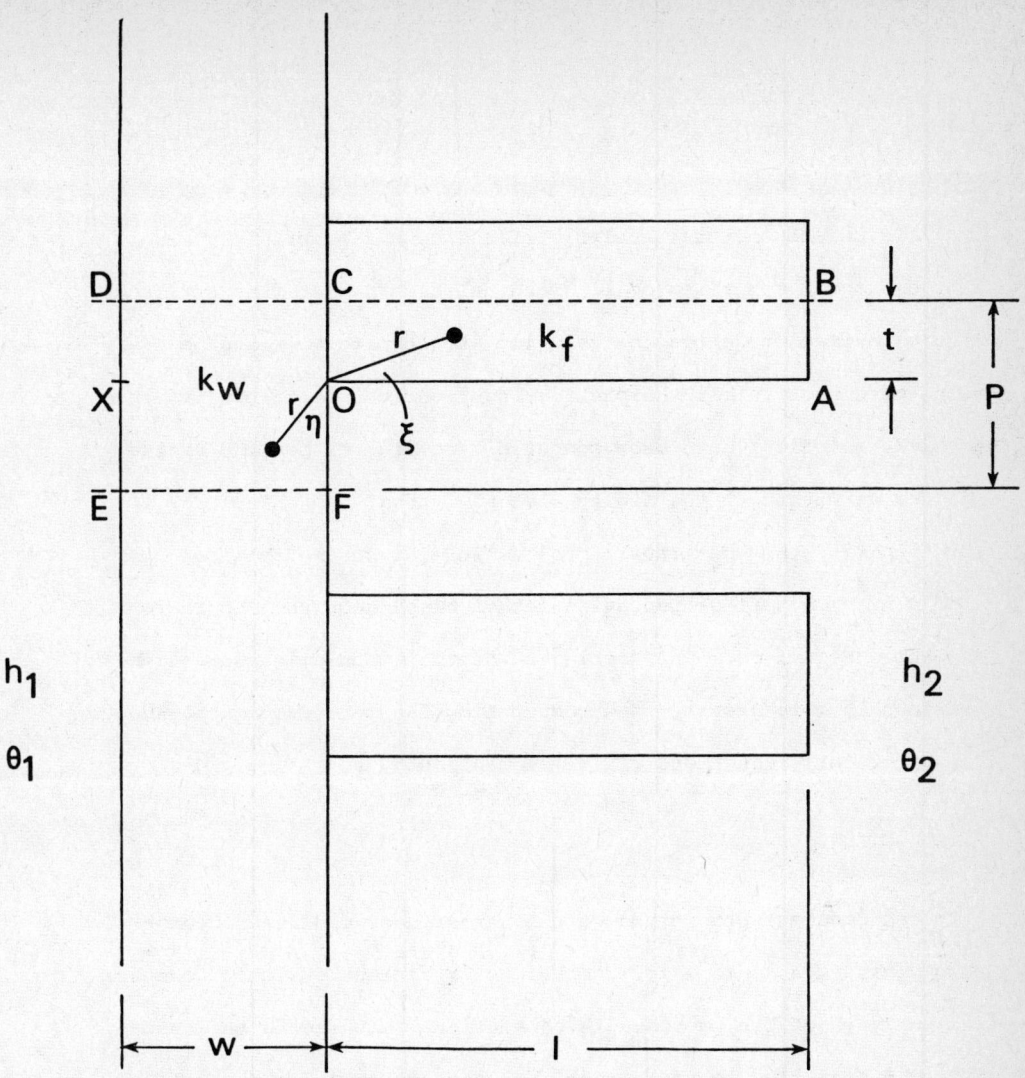

Fig.1 Schematic representation of the fin assembly

3.4 THE ANALYSIS OF FIN ASSEMBLY HEAT TRANSFER BY A SERIES TRUNCATION METHOD

ABSTRACT

The two-dimensional analysis of the heat flow within a finned heat exchanger comprised of longitudinal rectangular fins attached to a plane wall requires the solution of a Laplacian mixed boundary-value problem. In all the previous publications pertaining to this problem, solutions have been computed employing either the finite-difference method or the finite-element method. In this study it is shown that this particular problem is also susceptible to treatment by a series truncation method based upon the separation of variables technique. This series truncation method is shown to be numerically more accurate and computationally more economical than the finite-difference and finite-element methods.

INTRODUCTION

The accurate prediction of the performance of finned heat exchangers is essential for compact and efficient design. However, the analysis of fin problems is conventionally based on the assumption that the heat flow is uni-directional because this, in general, facilitates an analytical treatment, e.g. [1,2]. The early investigations into the applicability of the one-dimensional approximation restricted attention solely to the fin and concluded that two-dimensional effects are negligible provided the transverse Biot number, based on the fin-base thickness, is much less than unity [3,4,5]. Recent investigations of the combined fin and supporting surface have shown that the presence of fins induces two-dimensional effects within the supporting surface and these may in turn act to produce two-dimensional variations within the fin, e.g. [6,7,8,9]. Suryanarayana [7] has reported that the difference between heat transfer rates predicted by one- and two-dimensional analyses can be as much as 80 per cent. It is therefore essential for the effective design of finned heat exchangers to consider the complete fin assembly and to employ a multi-dimensional analysis.

The two-dimensional analysis of the conductive-convective heat flow through an assembly of longitudinal rectangular fins attached to a plane wall, Fig. 1, requires the solution of a Laplacian mixed boundary-value problem [7,8,9]. In the previous examinations of this problem solutions have been computed employing either the finite-difference method [7,8,9] or the finite-element method [9]. However, these methods necessitate the solution of very large systems of simultaneous linear algebraic equations in order to produce accurate solutions. Furthermore, Stones [9] has shown that, for a particular range of the system parameters, neither of these methods produces satisfactory solutions.

In this study it is shown that, contrary to the inferences of the previous investigations [7,8,9], this particular fin assembly problem is susceptible to an analytical treatment. An extension of the separation of variables technique is devised which enables the derivation of a truncated series solution. However, this solution is not in a closed form. The determination of the coefficients associated with the series expansions requires the solution of a system of linear algebraic equations. Nevertheless, in order to obtain solutions of comparable accuracy, the series truncation method requires the solution of considerably fewer equations than that required by either the finite-difference or the finite-element methods. Thus, the series truncation method facilitates a considerable reduction in the computational storage and time requirements. Furthermore, it is shown that the series truncation method yields accurate solutions even for problems for which the finite-difference and finite-element methods fail to provide acceptable results.

ANALYSIS

The following analysis is based upon the classical assumptions employed in the examination of conducting-convecting finned surfaces, namely, constant thermal conductivities, uniform heat transfer coefficients and perfect wall-to-fin contact.

The geometrical symmetry of an assembly of equally-spaced longitudinal rectangular fins attached to a plane wall indicates that it is only necessary to examine that section of the assembly shown schematically in Fig. 1. Thus, for steady-state two-dimensional conductive heat flow, the determination of the fin assembly temperature distribution, $\phi(X,Y)$, involves the simultaneous solution of [8],

$$\frac{\partial^2}{\partial X^2} \phi_f + \frac{\partial^2}{\partial Y^2} \phi_f = 0 \quad \text{within the fin,} \quad (1)$$

and

$$\frac{\partial^2}{\partial X^2} \phi_w + \frac{\partial^2}{\partial Y^2} \phi_w = 0 \quad \text{within the wall,} \quad (2)$$

subject to the boundary conditions [8],

on OA $\quad \dfrac{\partial \phi_f}{\partial Y} = 0$ (3a)

on AB $\quad \dfrac{\partial \phi_f}{\partial X} = - \dfrac{Bi_2}{\kappa} \phi_f$ (3b)

on BC $\quad \dfrac{\partial \phi_f}{\partial Y} = - \dfrac{Bi_2}{\kappa} \phi_f$ (3c)

on CO $\quad \phi_f = \phi_w$ (3d)

and $\quad \dfrac{\partial \phi_w}{\partial X} = \kappa \dfrac{\partial \phi_f}{\partial X}$ (3e)

on CD $\quad \dfrac{\partial \phi_w}{\partial X} = - Bi_2 \phi_w$ (3f)

on DE $\quad \dfrac{\partial \phi_w}{\partial Y} = 0$ (3g)

on EF $\quad \dfrac{\partial \phi_w}{\partial X} = - Bi_1 (1-\phi_w)$ (3h)

on FO $\quad \dfrac{\partial \phi_w}{\partial Y} = 0$ (3i)

It can be deduced from the equations (1), (2) and (3) that the heat flow within the fin assembly may be parameterised by the Biot numbers, Bi_1 and Bi_2, the ratio of the thermal conductivities κ, and the aspect ratios, L, T and W.

Integrating the equations (1) and (2) by the application of the separation of variables method [10] and then enforcing the boundary conditions (3a), (3b), (3c), (3g), (3h) and (3i) gives

$$\phi_f(X,Y) = \sum_{n=1}^{\infty} \dfrac{a_n (\cosh\lambda_n (L-X) + \dfrac{Bi_2}{\kappa \lambda_n} \sinh\lambda_n (L-X)) \cos\lambda_n Y}{\cosh\lambda_n L + \dfrac{Bi_2}{\kappa \lambda_n} \sinh\lambda_n L} \qquad (4)$$

and

$$\phi_w(X,Y) = 1 + b_o(1 + Bi_1(W + X)) + \sum_{n=1}^{\infty} \dfrac{b_n (\cosh\mu_n (W + X)) + \dfrac{Bi_1}{\mu_n} \sinh\mu_n (W+X)) \cos\mu_n Y}{\cosh\mu_n W + \dfrac{Bi_1}{\mu_n} \sinh\mu_n W}$$

(5)

where the eigenvalues λ_n and μ_n are defined by

$$\lambda_n \tan\lambda_n T = Bi_2/\kappa$$

and

$$\mu_n = n\pi$$

and the constant coefficients a_n and b_n are, as yet, undetermined.

Enforcing the remaining boundary conditions, (3d), (3e), and (3f), gives

$$0 = 1 - \sum_{n=1}^{\infty} a_n \cos\lambda_n Y + b_o(1 + Bi_1 W) + \sum_{n=1}^{\infty} b_n \cos\mu_n Y, \quad 0 \leq Y \leq T \tag{6}$$

$$0 = \kappa \sum_{n=1}^{\infty} a_n L_n \cos\lambda_n Y + b_o Bi_1 + \sum_{n=1}^{\infty} b_n M_n \cos\mu_n Y, \quad 0 \leq Y \leq T \tag{7}$$

and

$$0 = Bi_2 + b_o(Bi_1 + Bi_2(1 + Bi_1 W)) + \sum_{n=1}^{\infty} b_n(Bi_2 + M_n)\cos\mu_n Y, \quad T \leq Y \leq 1 \tag{8}$$

where

$$L_n = \lambda_n \left(\tanh\lambda_n L + \frac{Bi_2}{\kappa\lambda_n}\right) / \left(1 + \frac{Bi_2}{\kappa\lambda_n}\tanh\lambda_n L\right) \tag{9}$$

and

$$M_n = \mu_n \left(\tanh\mu_n W + \frac{Bi_1}{\mu_n}\right) / \left(1 + \frac{Bi_1}{\mu_n}\tanh\mu_n W\right) \tag{10}$$

Multiplying equation (7) by $\cos\lambda_m Y$ and then integrating over the range $0 \leq Y \leq T$ defines a relation between the a_n and b_n, namely,

$$a_m = \frac{-b_o Bi_1 \frac{\sin\lambda_m T}{\lambda_m} - \sum_{n=1}^{\infty} b_n M_n \int_0^T \cos\lambda_m Y \cos\mu_n Y \, dY}{\frac{\kappa}{4\mu_m}(2\mu_m T + \sin 2\mu_m T)L_m}, \quad m = 1, 2, \ldots \tag{11}$$

Thus, it only remains to determine the coefficients b_n. However, the combined complexity of the governing relations, namely equations (6),

(8) and (11), precludes the possibility of obtaining an explicit expression for the coefficients b_n. In fact, these coefficients can only be determined approximately. The temperature distributions $\phi_f(X,Y)$ and $\phi_w(X,Y)$ are approximated by the first N terms in each of the series expansions (4) and (5), respectively, and the relations (6) and (8) are accordingly modified to

$$0 = 1 + \sum_{n=1}^{N} \frac{(b_o^* Bi_1 \frac{\sin\lambda_n T}{\lambda_n} - \sum_{m=1}^{N} b_m^* M_m \int_o^T \cos\lambda_n Y \cos\mu_m Y dY) \cos\lambda_n Y}{\frac{K}{4\mu_n}(2\mu_n T + \sin 2\mu_n T) L_n}$$

$$+ b_o^*(1 + Bi_1 W) + \sum_{n=1}^{N} b_n^* \cos\mu_n Y, \quad 0 \leq Y \leq T \quad (12)$$

and

$$0 = Bi_2 + b_o^*(Bi_1 + Bi_2(1 + Bi_1 W)) + \sum_{n=1}^{N} b_n^*(Bi_2 + M_n) \cos\mu_n Y,$$

$$T \leq Y \leq 1 \quad (13)$$

where the coefficients a_n^* have been replaced using the approximate form of equation (11), namely,

$$a_m^* = \frac{-b_o^* Bi_1 \frac{\sin\lambda_m T}{\lambda_m} - \sum_{n=1}^{N} b_n^* M_n \int_o^T \cos\lambda_m Y \cos\mu_n Y dY}{\frac{K}{4\lambda_m}(2\lambda_m T + \sin 2\lambda_m T) L_m}, \quad m = 1, 2, \ldots, N$$

and the asterisk denotes the approximate value of the respective quantity.

In order to determine the N+1 unknown coefficients, $b_o^*, b_1^*, \ldots, b_N^*$, and therefore the solution, a system of linear algebraic equations is generated from the expressions (12) and (13). The equations (12) and (13) are multiplied by $\cos\mu_n Y$ and then equation (12) is integrated over the range $0 \leq Y \leq T$ and equation (13) is integrated over the range $T \leq Y \leq 1$. Combining the integrated relations and then collocating for $m=1,2,\ldots,N$ generates a system of N linear algebraic equations in the N+1 unknowns b_n^*. One additional equation is obtained by integrating equation (12) over the range $0 \leq Y \leq T$ and equation (13) over the range $T \leq Y \leq 1$. The resulting system of simultaneous linear algebraic equations is solved employing a Gaussian elimination technique [10] and thus $\phi_f^*(X,Y)$ and $\phi_w^*(X,Y)$ are determined.

FIN ASSEMBLY HEAT TRANSFER RATE

The heat flow through the fin assembly is most conveniently expressed in the form of an augmentation factor, Aug, defined as the ratio of the heat transfer rate of the fin assembly to that of the unfinned wall operating under the same conditions [9]. Corresponding to the approximate solutions $\phi_f^*(X,Y)$ and $\phi_w^*(X,Y)$, this augmentation factor may be obtained by integrating the heat flux at the surface EF, and is given by,

$$\text{Aug}^* = -b_o^* \, \text{Bi}_1 \left[\frac{1}{\text{Bi}_1} + W + \frac{1}{\text{Bi}_2} \right] \qquad (14)$$

Thus, the series truncation method enables the fin assembly heat transfer rate to be evaluated without actually computing the temperature distribution within the fin assembly. In contrast, the finite-difference and finite-element formulations automatically generate the temperature distribution throughout the fin assembly section OABCDEFO.

RESULTS AND DISCUSSION

Solutions have been obtained for a wide range of the system parameters, Bi_1, Bi_2, κ, L, T and W. For each particular problem solutions were computed for the cases $N = 25, 50, 100$ and 200. In order to illustrate the performance of the series truncation method the results for three particular problems are presented here:

Problem	$N = 25$	Aug^* $N = 50$	$N = 100$	$N = 200$
A	5.1202	5.1167	5.1153	5.1149
B	1.0432	1.0433	1.0434	1.0434
C	1.1366	1.1371	1.1374	1.1375

These results correspond to the problems defined by:

Problem	Bi_1	Bi_2	κ	L	T	W
A	1.00	0.01	1.0	10.0	0.5	5.0
B	2.25	0.75	1.0	10.0	0.5	5.0
C	2.25	0.75	10.0	10.0	0.5	5.0

Problem A represents the performance of a stainless-steel finned heat exchanger with forced convection of water on the plain side and free convection of air on the fin side. The system parameters for the problems B and C lie in the range $Bi_1 > 2.0$ and $Bi_2 > 0.5$ for which Stones [9] has indicated that neither the finite-difference method nor the finite-element method provide acceptable solutions.

The results for the problems A, B and C illustrate various features of the series truncation technique and are characteristic of those observed for other values of the system parameters. In particular, in all cases, the augmentation factor, Aug*, displays a monotonic convergent behaviour as the order of the approximation is improved, i.e. as more terms are taken in the approximate representations of $\phi_f(X,Y)$ and $\phi_w(X,Y)$. Furthermore, the first three significant figures of the solution remain unchanged as the approximation is refined from $N = 50$ to $N = 200$.

Stones [9] performed a similar convergence study for the finite-difference and finite-element methods and concluded that even with as many as 991 codes, (requiring the solution of a system of 991 linear algebraic equations), these methods are usually only accurate to two significant figures. It is therefore apparent that the series truncation method facilitates substantial reductions in the computation requirements because for the case $N = 50$ the series truncation method only requires the solution of a system of 51 linear algebraic equations, but is accurate to at least three significant figures.

CONCLUSIONS

A mathematically rigorous method has been devised for the solution of a fin assembly heat transfer problem. This method uses the separation of variables technique in order to integrate the governing differential equations exactly, and approximations are only introduced in order to satisfy the boundary conditions pertaining to the wall-to-fin interface. The results obtained indicate that this method is more accurate and computationally more economical than the finite-difference and finite-element methods. Furthermore, this method gives accurate solutions even for problems for which the finite-difference and finite-element methods fail to provide acceptable results, e.g. see the results for problems B and C given in the preceeding section.

NOMENCLATURE

Aug	augmentation factor
Bi_1	$= h_1 P/k_w$, Biot number
Bi_2	$= h_2 P/k_w$, Biot number
h_1, h_2	heat transfer coefficients, $W/m^2 K$
k_f, k_w	thermal conductivities, W/mK
l	fin length, m
L	$= l/P$, aspect ratio
N	number of terms in truncation.
P	half fin pitch, m
t	half fin thickness, m
T	$= t/P$, aspect ratio
w	wall thickness, m
W	$= w/P$, aspect ratio
x	longitudinal displacement, m
X	$= x/P$, dimensionless longitudinal displacement
y	transverse displacement, m
Y	$= y/P$, dimensionless transverse displacement
κ	$= k_f/k_w$
θ	temperature distribution, K
θ_1, θ_2	fluid temperatures, K
ϕ	$= (\theta-\theta_2)/(\theta_1-\theta_2)$, dimensionless temperature distribution
*	denotes approximate value

Subscripts

1   plain side

2   fin side

f   fin

w   wall

REFERENCES

1. K.A. Gardner, "Efficiency of Extended Surface", Transactions of the ASME, Vol. 67, pp. 621-631, 1945.

2. I. Mikk, "Convective Fin of Minimum Mass", International Journal of Heat and Mass Transfer, Vol. 23, pp. 707-711, 1980.

3. R.K. Irey, "Errors in the One-Dimensional Fin Solution", Journal of Heat Transfer, Vol. 90, pp. 175-176, 1968.

4. M. Levitsky, "The Criterion for the Validity of the Fin Approximation", International Journal of Heat and Mass Transfer, Vol. 15, pp. 1960-1963, 1972.

5. W. Lau and C.W. Tan, "Errors in One-Dimensional Heat Transfer Analysis in Straight and Annular Fins", Journal of Heat Transfer, Vol. 95, pp. 549-551, 1973.

6. E.M. Sparrow and L. Lee, "Effects of Fin Base-Temperature Depression in a Multifin Array", Journal of Heat Transfer, Vol. 97, pp. 463-465, 1975.

7. N.V. Suryanarayana, "Two-Dimensional Effects on Heat Transfer Rates from an Array of Straight Fins", Journal of Heat Transfer, Vol. 99, pp. 129-132, 1977.

8. P.J. Heggs and P.R. Stones, "The Effects of Dimensions on the Heat Flow Rate through Extended Surfaces", Journal of Heat Transfer, Vol. 102, pp. 180-182, 1980.

9. P.R. Stones, Ph.D. Thesis, Leeds University, England, 1980.

10. A. Ralston, A First Course in Numerical Analysis, McGraw-Hill, New York, 1965.

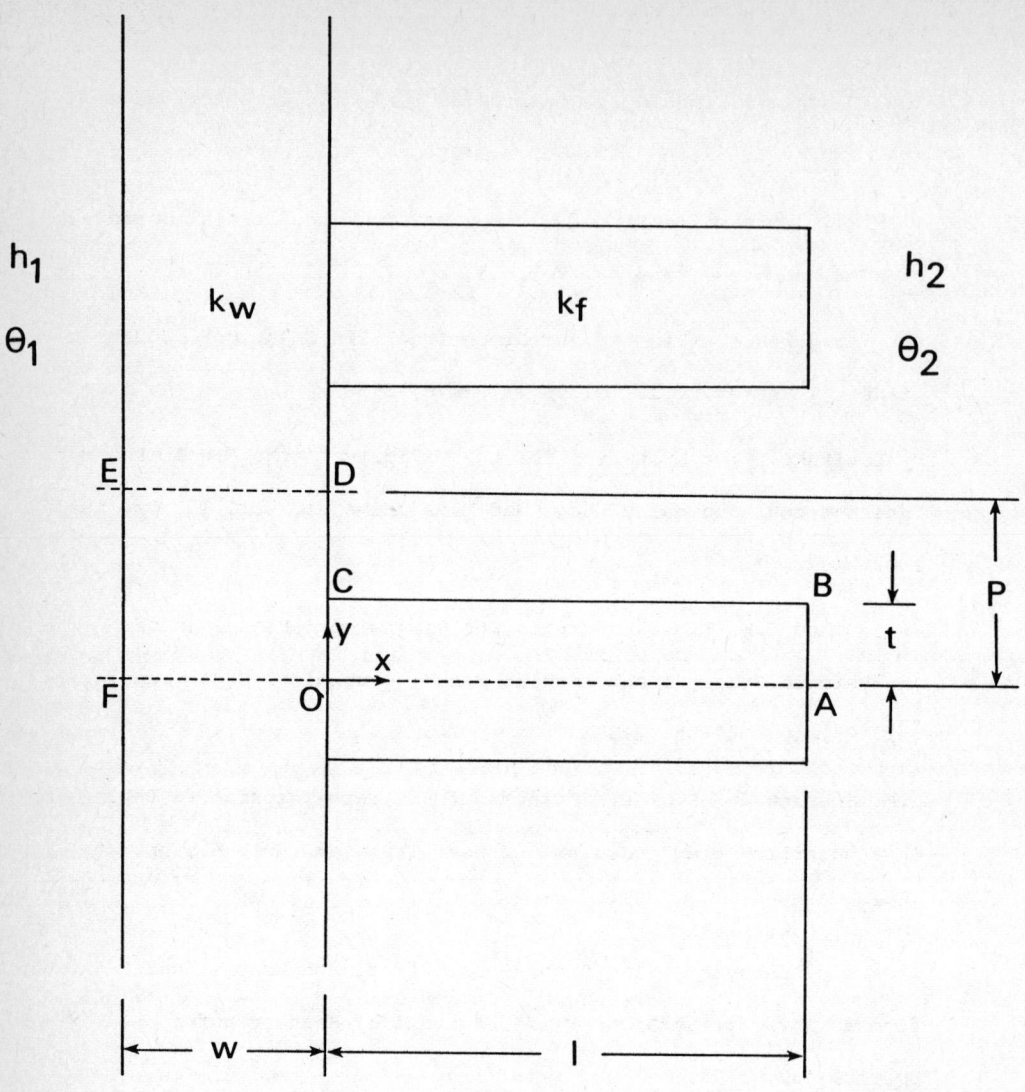

Fig.1 Schematic representation of the fin assembly

3.5 THE TWO-DIMENSIONAL ANALYSIS OF FIN ASSEMBLY HEAT TRANSFER:

A COMPARISON OF SOLUTION TECHNIQUES

ABSTRACT

The two-dimensional analysis of the heat flow within a finned heat exchanger which comprises longitudinal rectangular fins attached to a plane wall requires the solution of a Laplacian mixed boundary-value problem. This can be achieved by various numerical techniques e.g. the finite-difference, finite-element and boundary integral equation methods. In addition, this particular fin assembly problem is also susceptible to treatment by a series truncation method. In this study the relative performance of the finite-difference, finite-element, boundary integral equation and series truncation methods in determining the fin assembly heat transfer rate are investigated. The capabilities and limitations of each method are also discussed.

INTRODUCTION

The theoretical study of the heat flow within finned heat exchangers is of considerable practical importance because of the extensive utilisation of fins for heat transfer enhancement in applications varying from gas liquefaction plant to heat rejection equipment in motor vehicle engines. The accurate prediction of the thermal performance of finned heat exchangers is essential for compact and efficient design. However, in the analysis of such systems, it is conventionally assumed that the heat flow is one-dimensional because this, in general, facilitates an analytical treatment, e.g. [1,2]. The early investigations into the applicability of the one-dimensional approximation restricted attention solely to the fin and concluded that two-dimensional effects are negligible provided the transverse Biot number, based on the fin-base thickness, is much less than unity, e.g. [3,4,5]. Recent investigations of the combined fin and supporting surface have shown that the presence of fins induces two-dimensional effects within the supporting surface and these may in turn cause two-dimensional variations within the fin, e.g. [6,7,8]. Surayanarayana [7] has reported that the difference between fin assembly heat transfer rates predicted by one- and two-dimensional analyses can be as much as 80 percent. It is therefore essential for the effective design of finned heat exchangers to consider the complete fin assembly and to employ a multi-dimensional analysis.

The two-dimensional analysis of conductive-convective heat flow through an assembly of longitudinal rectangular fins attached to a plane wall requires the solution of a Laplacian mixed boundary-value problem, e.g. [6,7,8]. This can be achieved by various numerical techniques, e.g. the finite-difference [9], finite-element [10] and boundary integral equation [11] methods. This fin assembly problem is also susceptible to an analytical treatment [12]. The separation of variables method enables the derivation of a series solution. However, this solution is not in a closed form; the determination of the coefficients associated with the series expansion requires the solution of a system of linear algebraic equations. The generation of this algebraic representation is achieved by truncating the series and then collocating in a manner analogous with the discretization process of a numerical scheme.

The main objective of the present study is to investigate the relative performance of the finite-difference (FD), finite-element (FE), boundary integral equation (BIE) and series truncation (ST) techniques in determining the heat transfer rate through an assembly of longitudinal rectangular fins attached to a plane wall. In this investigation the general situation in which the fin and supporting surface have different thermal conductivities is considered.

ANALYSIS

The following analysis is based upon the classical assumptions employed in the examination of conducting-convecting finned surfaces, namely, constant thermal conductivities, uniform heat transfer coefficients and perfect wall-to-fin contact, e.g. [6,7,8]. These simplifications are introduced not only to reduce the complexity of the problem but also to minimise the number of variable parameters while still retaining the essential features of the actual physical situation.

Consider an assembly of longitudinal rectangular fins attached to a plane wall, as depicted schematically in Fig. 1. The geometrical symmetry of the fin arrangement indicates that it is only necessary to examine that section of the assembly bounded by the contour OABCDEFO, Fig. 1. Therefore, for steady-state two-dimensional heat flow, the determination of the fin assembly temperature distribution requires the simultaneous solution of [8],

$$\nabla^2 \phi_w = 0 \qquad \text{within } \Omega_w, \text{ (Fig. 1)}, \qquad (1)$$

and

$$\nabla^2 \phi_f = 0 \qquad \text{within } \Omega_f, \text{ (Fig. 1)}, \qquad (2)$$

subject to the boundary conditions

on OA $\qquad \phi_f' = - \dfrac{Bi_2}{\kappa} \phi_f \qquad \qquad$ (3a) on AB $\quad \phi_f' = -\dfrac{Bi_2}{\kappa} \phi_f$ (3b)

on BC $\quad \phi_f' = 0$ (3c)

on CO $\quad \phi_f = \phi_w$ (3d)

and $\quad \phi_w' = -\kappa \phi_f'$ (3e)

on CD $\quad \phi_w' = 0$ (3f)

on DE $\quad \phi_w' = Bi_1 (1 - \phi_w)$ (3g)

on EF $\quad \phi_w' = 0$ (3h)

on FO $\quad \phi_w' = -Bi_2 \phi_w$ (3i)

where the prime (') denotes the derivative in the direction of the outward normal to the associated surface.

Conditions (3c), (3f) and (3h) arise from the geometric and thermal symmetry of the fin assembly configuration and stipulate that there is no heat flux across the fictitious boundaries BC, CD and EF, respectively. Conditions (3d) and (3e) arise from the assumption of perfect wall-to-fin contact which requires that the temperature and heat flux be continuous across the contact interface, OC. The remaining boundary conditions describe the convective heat exchange from the exposed surfaces, DE and FOAB. The heat flow through the assembly is parameterised by the Biot numbers, Bi_1 and Bi_2, the ratio of the thermal conductivities κ, and the aspect ratios L, T and W.

SOLUTION METHODS

(a) Finite Difference Method

With the finite-difference formulation the problem described by equations (1), (2) and (3) is replaced by a system of linear algebraic equations in which the unknowns are the temperatures at discrete points within the fin assembly domain, [9]. This algebraic representation is generated by a relatively simple process. A rectangular mesh is super-imposed over the domain Ω and nodes are situated at the lattice points of this mesh. Then, at each internal node the differential equations (1) and (2) are replaced by the appropriate five-point central difference approximations and at each boundary node the differential operators in the respective boundary condition are approximated by the appropriate three-point central difference formula [9]. Assembling these discretized expressions produces a system of simultaneous linear algebraic equations involving the unknown nodal temperatures. In this study the solution to this system of equations is obtained employing a Gaussian elimination technique [9] which accounts for the banded structure of the equations and thereby facilitates substantial reductions in the computational storage and time requirements. The possibility of solving the algebraic equations by an iterative technique, namely the method of successive-over-relaxation [9] has also been investigated. It was found that the iterative process does not always converge.

(b) Finite Element Method

The finite-element method involves the reformulation of the problem defined by equations (1), (2) and (3) in terms of a pair of coupled variational statements [10]; one pertaining to the temperature distribution within the wall, and the other to that within the fin. The coupling is a consequence of the continuity conditions (3d) and (3e). The fin assembly temperature distribution is determined by minimising the surface integrals occurring in the variational statements. In order to evaluate these surface integrals the fin assembly domain Ω is subdivided into a finite number of triangular elements and the temperature is assumed to vary linearly within each element. The minimisation is then performed by integrating separately over each of the triangular elements. This procedure generates a system of linear algebraic equations in the unknown element corner temperatures. The relative performances of a Gaussian elimination technique and the method of successive-over-relaxation in obtaining the solution to this system of equations has been investigated. It was found that the method of successive-over-relaxation does not always give a convergent solution in that the iterative process fails to converge. Therefore, all the finite-element results presented in this study are computed employing a modified Gaussian elimination technique which takes into account the banded structure of the equations and thereby minimises the respective computational requirements.

(c) Boundary Integral Equation Method

The application of Green's Integral Formula [11] to the problem described by equations (1), (2) and (3) gives rise to a pair of coupled integral equations involving contour integrals around $\partial\Omega_f$ and $\partial\Omega_w$. The coupling arises from the interface boundary conditions (3d) and (3e). In order to obtain a solution to the integral equations, the contours

$\partial\Omega_f$ and $\partial\Omega_w$ are subdivided into rectilinear segments and nodes are situated at the midpoint of each of these segments. The temperature and heat flux on each segment are approximated by piecewise-constant functions. Then, the discretized form of the integral equations is collocated at each of the boundary nodes. This generates a system of linear algebraic equations involving the unknown nodal temperatures. This system of equations is considerably smaller than that generated by an equivalent finite-difference or finite-element representation because the BIE discretization occurs only on the boundary of the fin assembly domain Ω. Consequently, even for fine discretizations, it is most appropriate to solve these equations by a direct method such as Gaussian elimination. Nevertheless, the possibility of obtaining solutions employing the method of successive-over-relaxation has also been investigated. It was found that the iterative process fails to converge irrespective of the values of the system parameters, the size of the boundary discretization and the magnitude of the relaxation parameter.

(d) Series Truncation Method

The separation of variables technique enables the differential equations (1) and (2) to be integrated exactly and gives the solutions in terms of infinite series of hyperbolic and trignometric functions [12]. The coefficients and eigenvalues associated with these series expansions are determined from the boundary conditions (3). Enforcing the conditions (3b), (3c), (3f) and (3h) gives rise to expressions which explicitly define the eigenvalues. Enforcing the remaining boundary conditions generates a set of relations involving the unknown coefficients.

However, these relations are not in a closed form. Therefore, in order to determine the coefficients it is necessary to truncate the series and then to enforce these relations approximately. This generates a system of linear algebraic equations in which the unknowns are the coefficients associated with the truncated series. The solution to this system of equations is achieved by Gaussian elimination and the corresponding temperature distribution is obtained by appropriate summation of the series.

DISCUSSION OF SOLUTION METHODS

From the brief descriptions presented in the preceeding section it is apparent that there are basic conceptual differences between the various solution techniques. The FD formulation tackles the governing equations directly in the prescribed form, i.e. without any further mathematical analysis and requires discretization over the entire domain. The FE method reformulates the prescribed problem in terms of variational statements but nevertheless involves approximations throughout the fin assembly domain, Ω. The BIE method firstly transforms the prescribed problem into an equivalent set of integral equations by means of an analytical integration which, in effect, reduces the dimension of the problem by one. These integral equations only involve the boundary values of the temperature and consequently approximations are only introduced on the domain boundary, $\partial\Omega$. With the ST technique the governing equations are integrated exactly and approximations are only introduced in order to satisfy the boundary conditions (3d), (3e) and (3i). This ST technique is, in essence, equivalent to a numerical technique which uses very sophisticated polynomial approximations to the solution and applies these at every point within the fin assembly domain. In contrast, the FD, FE and BIE techniques use comparatively simple approximations and apply these only at discrete points within the domain.

In order to critically assess the relative performance of the solution techniques it is necessary to define some form of equivalence of the various forms of discretization. This is readily accomplished for the FD, FE and BIE methods because the application of each of these methods involves some form of subdivision of the fin assembly domain; the discretization of each of these methods is simply based upon a comparable subdivision of the domain. A rectangular grid is established over the domain Ω and the FD discretization is performed on this grid. Each of the rectangular sub-regions of the FD grid is subdivided along a diagonal, as shown in Fig. 2, and the resulting triangular elements are used for the FE discretization. The boundary elements of the FD grid are used as the boundary segments for the BIE discretization, Fig. 2. Thus, the FD, FE and BIE implementations are considered to be equivalent if the respective discretizations are based upon the same mesh size.

It is not possible to relate the ST discretization to that of the FD, FE and BIE methods because the ST solution procedure does not involve a subdivision of the fin assembly domain. However, as the algebraic representations generated by the BIE and ST methods are similarly dense, as opposed to the banded system of equations produced by the FD and FE formulations, the ST discretization shall be considered to be equivalent to that of the other techniques if the BIE and ST algebraic representations are of the same size, i.e. if the number of terms taken in the truncated series solution is equal to the number of boundary segments employed in the implementation of the BIE method.

HEAT EXCHANGER PERFORMANCE

The heat flow rate through the fin assembly is most conveniently expressed in the form of an augmentation factor, Aug, which is defined as the ratio of the heat transfer rate of the fin assembly to that of the unfinned wall operating under the same conditions [8]. This augmentation factor can be evaluated at either of the exposed surfaces DE and FOAB [8], and is given by

$$\text{Aug} = \left(\frac{1}{Bi_1} + W + \frac{1}{Bi_2}\right) \int_{DE} Bi_1 (1 - \phi_w(q)) dq \qquad (4)$$

$$= \left(\frac{1}{Bi_1} + W + \frac{1}{Bi_2}\right) \left\{ \int_{FO} Bi_2 \phi_w(q) dq + \int_{OAB} Bi_2 \phi_f(q) dq \right\} \qquad (5)$$

In the context of the FD, FE and BIE solutions these integrations can be performed employing an appropriate quadrature formula. However, as these techniques only provide approximate solutions, the corresponding values for the integrations (4) and (5) need not be exactly the same, although, for the solutions to be satisfactory these should agree to within an acceptable tolerance. Therefore, in the subsequent calculations Aug_1 and Aug_2 shall denote the values of the augmentation factor corresponding to the expressions (4) and (5) respectively. A further requirement for the solutions to be satisfactory is that the corresponding augmentation factors show a convergent behaviour as the order of the approximation is improved, i.e. as the mesh is refined, in the case of the FD, FE and BIE solutions, and as more terms are taken in the truncated series, in the case of the ST solutions.

With the FD and FE formulations the evaluation of the expressions (4) and (5) requires the determination of the temperature distribution throughout the domain Ω. In contrast, the BIE method yields all the necessary information for the computation of these quantities, namely the boundary distribution of ϕ, when the boundary integral equation representation of the problem described by equations (1), (2) and (3) is solved. With the BIE formulation the fin assembly temperature distribution is simply generated from that on the boundary and need only be computed if desired.

The ST technique enables the integrations (4) and (5) to be performed analytically and has the added advantage that these may be evaluated without computing the temperature at any point within the assembly [12].

RESULTS AND DISCUSSION

The solution to the problem described by equations (1), (2) and (3) by any technique which involves approximations in the solution procedure invariably includes an error. The errors in the solutions predicted by the FD, FE and BIE methods are related to the associated mesh sizes [9,10,11], whilst those in the ST solutions are dependent upon the length of the truncated series [12], e.g. the central difference approximations used in FD method introduce a discretization error of the $O(H^2)$, where H is the largest of the associated mesh spacings [9]. These errors diminish as the respective discretization is refined and consequently the approximate solutions approach the exact solution. In order to check for this convergence, solutions are computed for three different levels of discretization. These are denoted by discretization (A), discretization (B) and discretization (C) and, for the FD, FE and BIE methods, correspond to the FD grids described by

Discretization	H(OA)	H(OC)	H(OX)	H(OF)
(A)	OA/20	OC/5	OX/5	OF/5
(B)	OA/40	OC/10	OX/10	OF/10
(C)	OA/80	OC/20	OX/20	OF/20

These particular discretizations were found to offer the most efficient use of the computational resources with respect to the accuracy of the corresponding solutions. The discretizations (A), (B) and (C) correspond to a total of 186, 671 and 2541 nodes for each of the FD and FE methods, and 80, 160 and 320 boundary segments for the BIE method. The equivalent ST discretization take 80, 160 and 320 terms in the truncated series.

Solutions have been computed for a wide range of the system parameters Bi_1, Bi_2, κ, L, T and W. The results for 3 particular problems are presented in Tables 1, 2 and 3. These Tables show the values of the augmentation factors, Aug_1 and Aug_2, and the respective ratios of Aug_1 to Aug_2, as predicted by the FD, FE, BIE and ST methods. These results correspond to the problems defined by the system parameters,

1. $Bi_1 = 1.0$, $Bi_2 = 0.010$ $\kappa = 1.0$ $L = 5.0$ $T = 0.25$ and $W = 1.0$
2. $Bi_1 = 0.2$ $Bi_2 = 0.001$ $\kappa = 10.0$ $L = 2.0$ $T = 0.20$ and $W = 0.5$
3. $Bi_1 = 5.0$ $Bi_2 = 0.200$ $\kappa = 20.0$ $L = 4.0$ $T = 0.20$ and $W = 2.0$

The solutions to these problems illustrate various features of the FD, FE, BIE and ST solution techniques and are characteristics of the solutions observed for other values of the system parameters. In particular, for all problems considered, the solutions predicted by all four solution techniques display a convergent behaviour as the corresponding discretization is refined. However, these solutions invariably fail to achieve their respective limiting values. The primary cause of this slow convergence is the presence of a boundary singularity at the re-entrant corner, O, Fig. 1 [11]. Improvements in accuracy could be achieved by employing modified implementations which give special treatment to the singular point and thereby facilitate solutions which converge more rapidly, e.g. [13,14,15]. However, in comparison with the standard methods, these modified implementations necessitate considerably more programming effort and involve significant changes in the data structure, e.g. [13,14,15]. A more practical method for obtaining the limiting solutions would be to employ some form of extrapolation, e.g. for problems 1, 2 and 3 the limiting values of Aug_1 and Aug_2 have been computed using Richardson's Formula [16],

$$Err(N) \alpha (H(N))^\alpha \qquad (6)$$

where Err(N) is the error in the solution given by the N equation discretization, H(N) is an associated mesh size and α is the order of the extrapolation; these limiting solutions are included in Tables 1, 2 and 3. The excellent agreement between the extrapolated values of Aug_1 and Aug_2 emphasises the suitability of the Richardson's extrapolation method for obtaining the limiting values of the solutions.

A salient feature of all the results obtained, and clearly evident in Tables 1, 2 and 3, is the fact that the accuracy of the ST solutions is virtually independent of the system parameters; in all cases the first three significant figures in the ST solutions remain unchanged as the truncated series is extended from 80 to 320 terms, e.g. Tables 1, 2 and 3. Furthermore, although the rate of convergence of the

FD and FE methods varies from problem to problem, it has been found that the change in the values of Aug_1 and Aug_2 predicted by these methods, as the discretization is refined from case (A) to case (C), is always less than 5 per cent. The rate of convergence of the BIE solutions also changes from problem to problem, but the changes in the values of Aug_1 and Aug_2, as the discretization is refined from case (A) to case (C), are slightly greater than those evident in the FD, FE and ST solutions. In fact, it has been found that the BIE solutions can change by up to 15 per cent as the boundary discretization is refined from 80 to 320 boundary segments. Nevertheless, the extrapolated values of the BIE solutions are in excellent agreement with those of the FD and FE (and also ST) solutions. These variations in the rate of convergence of the FD, FE and BIE solutions can be attributed to the fact that the accuracy of the solutions predicted by these methods is dependent upon the mesh size used for discretization. The mesh size is, in turn, dependent upon the dimensions of the region Ω. The fin assembly dimensions for problem 2 are considerably smaller than those for problem 3, consequently, the errors in the solutions for problem 2 are less than those for problem 3.

The results predicted by the four solution techniques always show a convergent behaviour as the respective discretization is refined, however, only the FE and ST solutions satisfy the energy conservation criterion that Aug_1 be identically equal to Aug_2, i.e. E = 1.0. The accuracy with which the FD solutions satisfy this condition improves as the mesh is refined from discretization (A) to discretization (C), although the condition E = 1.0 is never achieved by these solutions. However, the discretization (C) FD solutions are never more than 2 per cent in error of the requirement that E = 1. The accuracy with which the BIE solutions satisfy the energy conservation criterion also improves as the discretization is refined from case (A) to (C). However, it has been found that for discretization (A), the BIE solutions can result in values of E which differ by up to 40 per cent from the requirement that $E = 1.0$.

In order to assess the relative performance of the four solution techniques, in addition to comparing the accuracy of the solutions given by these methods, it is also necessary to compare the respective computational requirements. The computational storage and time requirements of the various implementations of the four solution techniques have therefore been determined and are displayed in Table 4. From the data presented in Table 4 it is clearly evident that for each particular level of discretization the ST method requires considerably more storage than the other methods, and furthermore, is substantially slower, except in comparison with the discretization (C) implementations of the FD and FE methods. However, whereas the discretization (A) ST solutions are always accurate to at least 3 significant figures, the discretization (C) FD, FE and BIE are usually only accurate to 2 significant figures. Thus, not only is the ST method more accurate than the other methods, but it requires substantially less computational storage and time in order to attain this higher degree of accuracy.

CONCLUSIONS

The relative performance of the FD, FE, BIE and ST methods in determining the heat transfer rate of an assembly of longitudinal rectangular fins attached to a plane wall has been investigated. The results indicate that the ST method is by far the best suited to this particular problem because it gives the most accurate solutions with the minimal computational requirements. However, this method is not applicable to problems involving curved or tapered fin profiles or non-uniform heat transfer coefficients. Therefore, on the basis of the present investigation, it is recommended that the BIE method be used for such problems because it gives solutions of comparable accuracy to the FD and FE methods, but is computationally more economical than these methods. Furthermore, the flexibility of the BIE discretization process is such that it can easily handle the complexities associated with curved boundaries and non-uniform heat transfer coefficients.

NOMENCLATURE

Aug	augmentation
Bi_1	$= h_1 P/k_w$, Biot number
Bi_2	$= h_2 P/k_w$, Biot number
E	$= Aug_1/Aug_2$
h	heat transfer coefficient, $W/m^2 K$
H(XY)	mesh spacing on boundary segment XY
k	thermal conductivity, W/mK
ℓ	fin length, m
L	$= \ell/P$, aspect ratio
P	half fin-pitch, m
t	half fin-base thickness, m
T	$= t/P$, aspect ratio
w	wall thickness, m
W	$= w/P$, aspect ratio
κ	$= k_f/k_w$
θ	temperature distribution, K
θ_1, θ_2	fluid temperature, K
ϕ	$= (\theta-\theta_2)/(\theta_1-\theta_2)$ dimensionless temperature
Ω_f	region bounded by OABCO
Ω_w	region bounded by OCDEFO
Ω	$= \Omega_f + \Omega_w$

$\partial\Omega_f$ contour OABCO

$\partial\Omega_w$ contour OCDEFO

$\partial\Omega$ $= \partial\Omega_f + \partial\Omega_w$

Subscripts

1 plain side

2 fin side

f fin

w wall

REFERENCES

1. Gardner, K.A., "Efficiency of extended surfaces", Transactions of the ASME., Vol. 67, pp. 621-631, 1945.

2. Mikk, I., "Convective fin of minimum mass", International Journal of Heat and Mass Transfer, Vol. 23, pp. 707-711, 1980.

3. Irey, R.K., "Errors in the one-dimensional fin solution", Journal of Heat Transfer, Vol. 90, pp. 175-176, 1968.

4. Levitsky, M., "The criterion for the validity of the fin approximation", International Journal of Heat and Mass Transfer, Vol. 15, pp. 1960-1963, 1972.

5. Lau, W. and Tan, C.W., "Errors in the one-dimensional heat transfer analysis", Journal of Heat Transfer, Vol. 95, pp. 549-551, 1973.

6. Sparrow, E.M. and Lee, L., "Effects of fin-base temperature depression in a multifin array", Journal of Heat Transfer, Vol. 97, pp. 463-465, 1975.

7. Suryanarayana, N.V., "Two-dimensional effects on heat transfer rates from an array of straight fins", Journal of Heat Transfer, Vol. 99, pp. 129-132, 1977.

8. Heggs, P.J. and Stones, P.R., "The effects of dimensions on the heat flow rate through extended surfaces", Journal of Heat Transfer, Vol. 102, pp. 180-182, 1980.

9. Smith, G.D., Numerical solution of partial differential equations, Oxford University Press, 1974.

10. Zienkiewicz, O.C., The finite element method in engineering, McGraw-Hill, London, 1971.

11. Jawson, M.A. and Symm, G.T., Integral equation methods in potential theory and electrostatics, Academic Press, London, 1977.

12. Heggs, P.J., Ingham, D.B. and Manzor, M., "The analysis of fin assembly heat transfer by a series truncation method", submitted for publication to the Journal of Heat Transfer, 1981.

13. Griffiths, D.F., "A numerical study of a singular elliptic boundary value problem", JIMA, Vol. 19, pp. 59-69, 1977.

14. Wait, R., "Singular isoparametric finite elements", JIMA, Vol. 20, pp. 133-141, 1977.

15. Symm, G.T., "Treatment of singularities in the solution of Laplace's equation, by an integral equation method", National Physical Laboratory Report NAC 31, 1973.

16. Ralston, A., A first course in numerical analysis, McGraw-Hill, New York, 1965.

Table 1 : The solutions for problem 1

METHOD	Discretization (A)			Discretization (B)			Discretization (C)			Extrapolated Limit		
	Aug_1	Aug_2	E	Aug_1	Aug_2	E	Aug_1	Aug_2	E	Aug_1	Aug_2	E
FD	4.509	4.285	1.052	4.389	4.278	1.026	4.331	4.275	1.013	4.278	4.274	1.001
FE	4.286	4.286	1.000	4.279	4.279	1.000	4.276	4.276	1.000	4.274	4.274	1.000
BIE	3.971	4.459	0.891	4.157	4.344	0.957	4.231	4.299	0.984	4.279	4.270	1.002
ST	4.275	4.275	1.000	4.274	4.274	1.000	4.273	4.273	1.000	4.273	4.273	1.000

Table 2 : The solutions for problem 2

METHOD	Discretization (A)			Discretization (B)			Discretization (C)			Extrapolated Limit		
	Aug_1	Aug_2	E	Aug_1	Aug_2	E	Aug_1	Aug_2	E	Aug_1	Aug_2	E
FD	3.086	2.963	1.041	3.024	2.963	1.021	2.996	2.963	1.011	2.963	2.963	1.000
FE	2.966	2.966	1.000	2.963	2.963	1.000	2.963	2.963	1.000	2.963	2.963	1.000
BIE	3.014	2.962	1.012	2.984	2.962	1.007	2.972	2.963	1.003	2.963	2.963	1.000
ST	2.964	2.964	1.000	2.963	2.963	1.000	2.963	2.963	1.000	2.963	2.963	1.000

Table 3 : The solutions for problem 3

METHOD	Discretization (A)			Discretization (B)			Discretization (C)			Extrapolated Limit		
	Aug_1	Aug_2	E	Aug_1	Aug_2	E	Aug_1	Aug_2	E	Aug_1	Aug_2	E
FD	2.038	1.959	1.040	1.969	1.931	1.020	1.935	1.916	1.010	1.901	1.900	1.000
FE	1.959	1.959	1.000	1.930	1.930	1.000	1.915	1.915	1.000	1.900	1.900	1.000
BIE	1.684	2.478	0.679	1.813	2.122	0.854	1.866	1.984	0.941	1.902	1.896	1.003
ST	1.894	1.894	1.000	1.896	1.896	1.000	1.897	1.897	1.000	1.899	1.899	1.000

Table 4 : Computational requirements

METHOD	Discretization (A)		Discretization (B)		Discretization (C)	
	Storage	CPU Time	Storage	CPU Time	Storage	CPU Time
FD	155	0.35	720	11.5	5400	263.0
FE	155	0.35	720	11.5	5400	263.0
BIE	125	1.15	320	4.5	1050	27.0
ST	195	3.00	575	17.0	2060	104.0

Storage in KBytes, and CPU Time in seconds

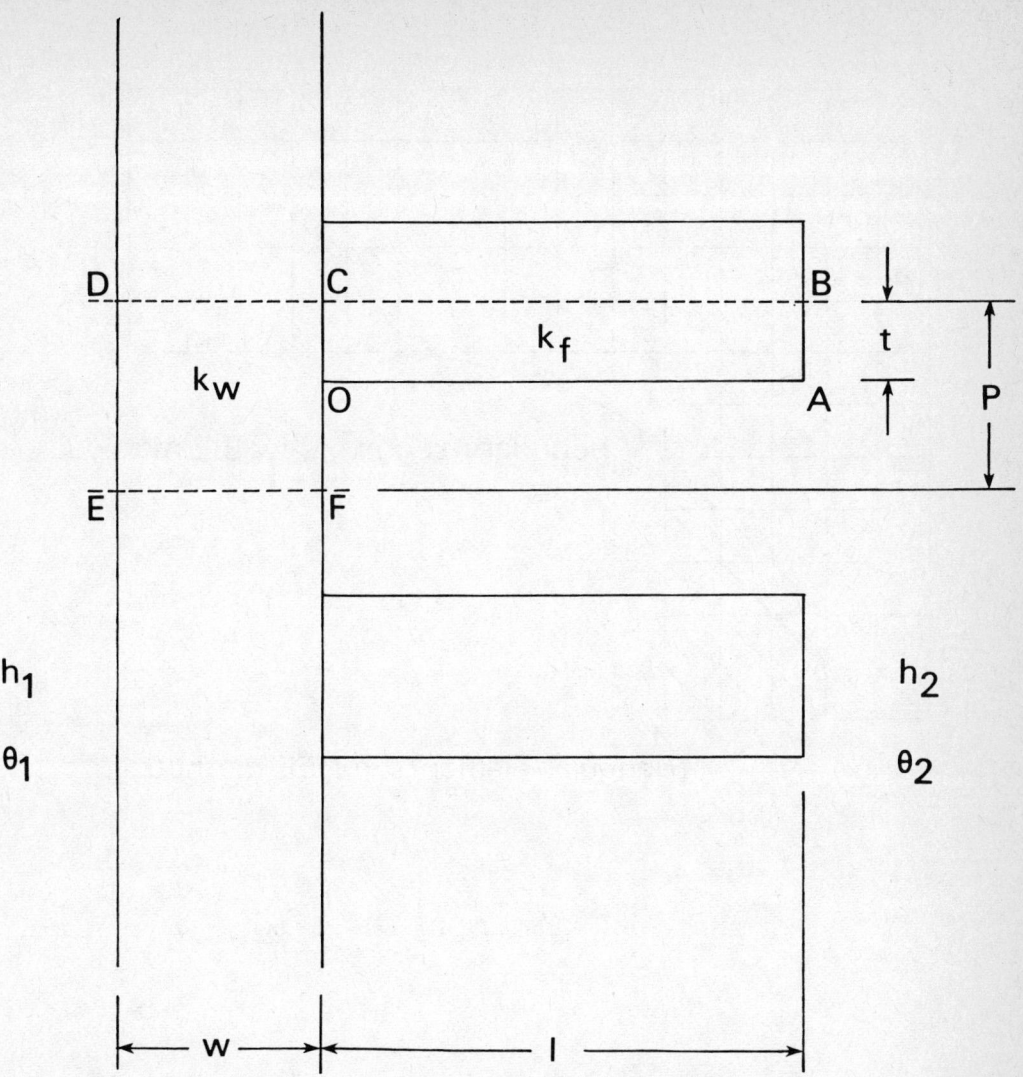

Fig.1 Schematic representation of the fin assembly

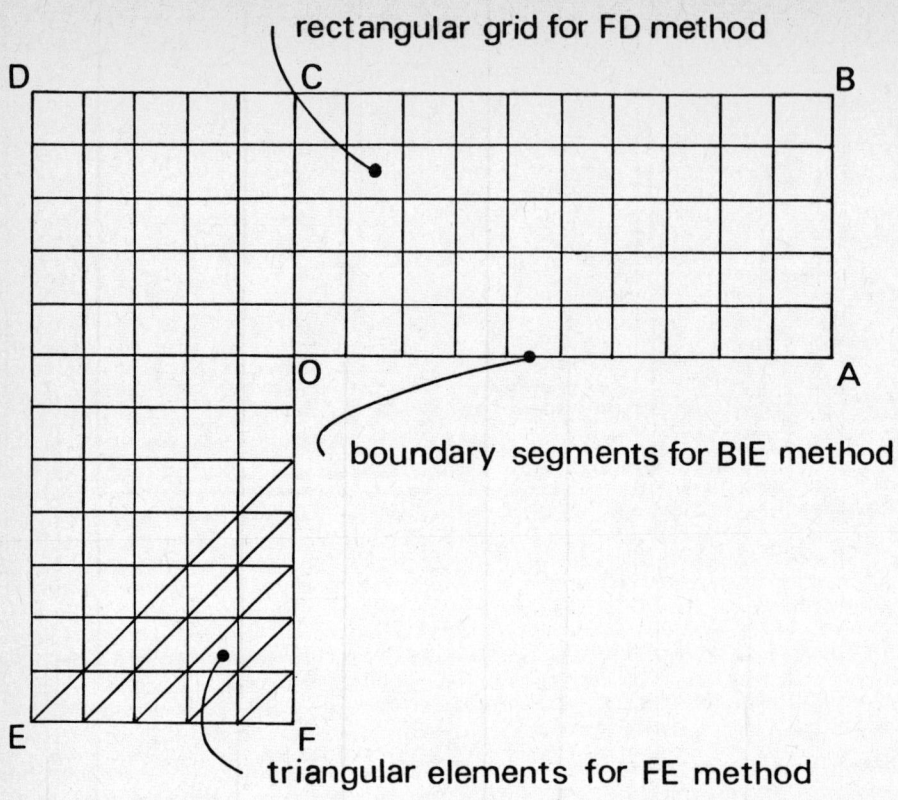

Fig.2 FD, FE and BIE discretizations

CHAPTER 4

THE ANALYSIS OF FIN RADIATION

4.1 THE BOUNDARY INTEGRAL EQUATION ANALYSIS OF NON-LINEAR PLANE POTENTIAL PROBLEMS

ABSTRACT

This paper presents the boundary integral equation (BIE) formulation and numerical solution procedure for two-dimensional problems governed by Laplace's equation and subject to non-linear boundary conditions. The introduction of non-linear terms constitutes a fundamental extension of the BIE method, as previous applications have been restricted entirely to linear problems. Furthermore, non-linearities necessitate the use of iterative solution techniques which present the conceptual disadvantage that a solution is not guaranteed. However, no difficulties were encountered with the Newton-Raphson iterative method employed in this study. The various features of the non-linear BIE formulation are illustrated by the application to a physical problem of relevance in heat exchanger design.

INTRODUCTION

Elliptic boundary value problems encountered in engineering and mathematical physics are generally intractable by analytical treatment. However, the solution to such problems can usually be computed employing numerical techniques such as the boundary integral equation [1], finite-difference [2] and finite-element [3] methods. The boundary integral equation (BIE) method is of less general applicability than the finite-difference (FD) and finite-element (FE) methods, but offers several important computational advantages over these methods for the solution of potential and biharmonic mixed boundary-value problems, e.g. [4-13]. The principal advantage of the BIE method is that, in contrast to the FD and FE methods, discretization for numerical purposes occurs only on the boundary of the relevant domain and therefore results in a considerably smaller system of equations than that generated by an equivalent FD or FE approximation. Furthermore, the BIE method can provide effective treatment of exterior or infinite domain problems, e.g. [1,11], for which the FD and FE methods are difficult to apply.

The FD and FE methods feature several advantageous properties besides their generality. In particular, the inherent necessity to discretize over the whole of the solution domain enables more efficient treatment of problems involving regional or material inhomogeneities, than possible with the BIE method. Furthermore, application of the FD and FE methods to non-linear problems is performed in much the same manner as for linear problems, although solution obviously involves greater algebraic complexity. In contrast, the BIE method is difficult to apply to problems which are governed by non-linear equations. However, it is shown in this study that the BIE method can be applied without difficulty to problems governed by linear equations but subject to non-linear boundary conditions. The BIE formulation and numerical solution procedure are described for plane Laplacian problems involving mixed linear and non-linear boundary conditions. The solution capabilities of this non-linear BIE method are illustrated by the application to a heat diffusion problem.

The inclusion of non-linear terms constitutes a fundamental extension of the BIE method, but has the inherent disadvantage that iterative techniques are introduced into the solution procedure; iterative techniques have the major deficiency that convergence, and therefore solution is not guaranteed. However, the Newton-Raphson iterative method [14] presented no such difficulties when applied to the problems investigated in this study.

FORMULATION AND NUMERICAL PROCEDURE

The BIE method is based upon Green's Integral Formula [1] which, for any sufficiently smooth function ϕ which satisfies Laplace's equation within a plane domain Ω having a piecewise-smooth boundary $\partial \Omega$, may be expressed as,

$$\int_{\partial \Omega} \{\phi(q) \log'|p-q| - \phi'(q) \log|p-q|\} dq = \eta(p)\phi(p) \qquad (1)$$

where,

i. $p \in \Omega + \partial\Omega$, $q \in \partial\Omega$

ii. dq denotes the differential increment of $\partial\Omega$ at q

iii. the prime ' denotes the derivative in the direction of the outward normal to $\partial\Omega$ at q, and

iv. if $p \in \Omega$ then $\eta = 2\pi$, but if $p \in \partial\Omega$ then η is the angle included between the tangents to $\partial\Omega$ on either side of p.

If either ϕ or ϕ' is prescribed at each point of $\partial\Omega$ then solution of the integral equation,

$$\int_{\partial \Omega} \{\phi(q) \log'|\bar{q}-q| - \phi'(q) \log|\bar{q}-q|\} dq - \eta(\bar{q})\phi(\bar{q}) = 0, \quad q,\bar{q} \in \partial\Omega \qquad (2)$$

determines the boundary distribution of both ϕ and ϕ'. The potential ϕ at any interior point can then be computed employing Green's Integral Formula, equation (1). Thus, application of Green's Boundary Formula, equation (2), enables two-dimensional Laplacian problems involving boundary conditions which prescribe either the potential ϕ or its normal derivative ϕ' at each point of the boundary, to be reformulated as integral equations in which the unknowns are the boundary values of ϕ and ϕ' complementary to those prescribed by the boundary conditions. However, if on some section of $\partial\Omega$ the boundary conditions specify neither ϕ nor ϕ', but prescribe a relation of the form $\phi' = f(\phi)$ then in equation (2) ϕ' is appropriately replaced by $f(\phi)$ and ϕ is treated as the relevent unknown.

Previous applications of the BIE method have been restricted entirely to problems in which the boundary conditions specify ϕ, ϕ' or a linear relation between ϕ and ϕ', e.g. [4-11]. The present study extends the application of the BIE method to problems which also involve non-linear boundary conditions. In order to facilitate concise explanation of the numerical solution procedure, attention is restricted to non-linearities of the form,

$$\phi'(q) = \alpha(q) + \sum_{i=1}^{r(q)} \beta_i(q)\, \phi^{\gamma_i(q)}(q), \quad q \in \partial\Omega \tag{3}$$

where α, β_i, γ_i and r are given functions. However, equation (3) includes the most commonly specified non-linear boundary conditions for physical problems.

In practice the integral equation (2) can rarely be solved analytically. Consequently, various numerical techniques have been devised for computing the solution. In this study, the classical or pieceswise constant BIE discretization [10] is employed. In the context of the class of non-linear boundary conditions defined by equation (3), the introduction of piecewise-linear [10] and piecewise-quadratic [10] boundary variations only involves a trivial extension of the present work.

In the classical BIE method the boundary $\partial\Omega$ is sub-divided into N smooth segments, $\partial\Omega_j$, j=1,....,N, on which ϕ and ϕ' are approximated by piecewise-constant functions ϕ_j and ϕ'_j, respectively. Correspondingly, Green's Integral Formula, equation (1), becomes,

$$\sum_{j=1}^{N} \{\phi_j \int_{\partial\Omega_j} \log'|p-q|dq - \phi'_j \int_{\partial\Omega_j} \log|p-q|dq\} = \eta(p)\phi(p), \quad p \in \Omega + \partial\Omega, \, q \in \partial\Omega \tag{4}$$

The application of this discretized form of Green's Integral Formula to the midpoint q_j of each segment generates a system of algebraic equations of the form,

$$A\Phi + B\Phi' = 0 \tag{5}$$

where,

$$A = (A_{ij}); \quad A_{ij} = \int_{\partial\Omega_j} \log'|q_i-q|dq - \delta_{ij}\eta(q_i) \tag{6}$$

$$B = (B_{ij}); \quad B_{ij} = \int_{\partial\Omega_j} \log|q_i-q|dq, \tag{7}$$

$\Phi = (\phi_1,...,\phi_N)^T$, $\Phi' = (\phi'_1,...,\phi'_N)^T$ and δ_{ij} denotes the Kronecker delta function. The subsequent introduction of the boundary conditions into the equation (5) yields a system of N simultaneous algebraic equations of the form,

$$F_i(x) = b_i, \quad i = 1,2,...,N \tag{8}$$

where $x = (x_1, \ldots, x_N)^T$ are the N unknown ϕ_j and ϕ_j' complementary to those prescribed and (b_i) are constants. If the boundary conditions include non-linearities of the form (3) then the system of equations (8) may be re-expressed as,

$$\sum_{j=1}^{N} A_{ij}^* x_j + \sum_{k} \sum_{j=1}^{N} (B_k^*)_{ij} x_j^{\gamma_k} = b_i, \quad i=1,\ldots,N, \tag{9}$$

where A_{ij}^* and $(B_k^*)_{ij}$ are appropriate combinations of A_{ij} and B_{ij}, equations (5), and γ_k are the various exponents in (3).

In order to compute the solution of a system of non-linear algebraic equations, such as that described by the expression (9), it is necessary to employ an iterative technique. From the expression (9) it can be deduced that the differentiations for the elements of the Jacobian matrix J [14] can be performed explicitly;

$$J_{ij}(x) = \frac{\partial}{\partial x} F_i(x) = A_{ij}^* + \sum_{k} (B_k^*)_{ij} \gamma_k x_j^{\gamma_k - 1} \tag{10}$$

Thus, the Newton-Raphson iterative method [14] appears most appropriate. The Newton-Raphson iteration determines the correction $\Delta x = (\Delta x_1, \ldots, \Delta x_N)^T$ to an iterate $x^{(n)}$ by solving the linearised equations,

$$\{F_i(x^{(n)}) - b_i\} + \sum_{j=1}^{N} \frac{\partial}{\partial x_j} F_i(x^{(n)}) \Delta x_j = 0, \quad i=1,\ldots,N, \tag{11}$$

The convergence of such a technique is not guaranteed, even if the iterative procedure is initiated with a close approximation to the desired root. However, assuming the iterative procedure does converge, the unknown ϕ_j and ϕ_j' are determined. Thus, knowing both ϕ_j and ϕ_j' at each point q_j, the potential at any point $p \in \Omega + \partial\Omega$ can be computed by a relatively simple quadrature, equation (4).

EXAMPLE PROBLEM

In order to demonstrate the method of application and illustrate the solution capabilities of the non-linear BIE formulation, a problem involving mixed linear and non-linear boundary conditions is considered. This problem arises in the study of heat transfer from finned surfaces, e.g. [15-19], and requires the determination of the steady-state temperature distribution within a rectangular fin which dissipates heat by both convection and radiation.

If the fin material is isotropic, obeys Fourier's law of heat conduction and contains no heat sources then, by symmetry, the temperature distribution within the fin satisfies,

$$\nabla^2 \phi = 0 \quad \text{within ABCDA, Fig. 1} \tag{12}$$

subject to the boundary conditions,

$$\text{on AB} \quad \phi' = (Bi\,\phi + Nc\,\phi^4) \tag{12a}$$

$$\text{on BC} \quad \phi' = -(Bi\,\phi + Nc\,\phi^4) \tag{12b}$$

$$\text{on CD} \quad \phi' = 0 \tag{12c}$$

$$\text{on DA} \quad \phi = 1 \tag{12d}$$

where Bi is a Biot number arising from the assumption that convective heat transfer satisfies Newton's law of cooling and Nc is a dimensionless combination of the fin-base temperature, the fin length (AB), the Stefan-Boltzmann constant and the thermal conductivity of the fin.

The boundary conditions (12a) and (12b) describe the heat dissipation from the surfaces AB and BC, respectively, and the boundary condition (12c) arises from the geometrical symmetry of the fin configuration. The boundary (12d) stipulates that the fin is attached to an isothermal surface.

The theoretical study of the heat flow within fins is of considerable practical importance because of the extensive utilisation of finned surfaces for heat transfer enhancement in applications varying from gas liquefaction plant to heat rejection equipment in motor vehicle engines. Consequently, the analysis of fin heat transfer has received considerable attention e.g. [15-19]. However, previous investigations of combined convective and radiative heat dissipation from finned surfaces have neglected the transverse, (i.e. η - direction, Fig. 1) heat conduction within the fin. Thus, solution of the two-dimensional problem described by equations (12) should facilitate an improvement in the design of heat exchangers.

The boundary conditions (12a) and (12b) are of the class (3). Therefore, the application of the classical BIE method to the problem described by the equations (12) generates a system of non-linear algebraic equations of the form,

$$\sum_{j=1}^{N} A_{ij}^* x_j + \sum_{j=1}^{N} B_{ij}^* x_j^4 = b_i, \quad i = 1, \ldots, N \tag{13}$$

In order to compute a solution to this system of non-linear algebraic equations, the Newton-Raphson iterative solution process is initiated with the 'iterate' $x^{(o)}$ corresponding to the case where $\phi = 1$ throughout the fin. Subsequent iterates are computed from the relation,

$$x_i^{(n+1)} = x_i^{(n)} + \Delta x_i^{(n)}, \quad i = 1, \ldots, N$$

where the correction $\Delta x^{(n)}$ is obtained by solving,

$$\left\{ \sum_{j=1}^{N} A_{ij}^* x_j^{(n)} + \sum_{j=1}^{N} B_{ij}^* \left[x_j^{(n)} \right]^4 - b_i \right\}$$

$$+ \left\{ \sum_{j=1}^{N} \{ A_{ij}^* + 4 B_{ij}^* \left[x_j^{(n)} \right]^3 \} \Delta x_j^{(n)} \right\} = 0, \quad i = 1, \ldots, N.$$

Solution of the equations (13) by this iterative process determines the nodal distributions of both ϕ and ϕ'. The potential at any point within the domain may then by computed from the discretized form of Green's Integral Formula, equation (4).

One of the quantities of physical importance in this problem is the heat transfer rate of the fin Q which is given by,

$$Q = \int_{ABC} -\phi'(q) \, dq \qquad (14)$$

It is apparent from the expression (14) that the evaluation of Q only requires the boundary distribution of ϕ'; this is precisely the information obtained when the boundary integral equation representation of the problem defined by the equations (12) is solved. This feature of the BIE formulation also enables a check on whether the BIE approximation to the solution satisfies the energy conservation requirement that,

$$E = \int_{ABCDA} \phi'(q) \, dq = 0 \qquad (15)$$

Results have been obtained for a wide range of the system parameters Bi, Nc and AB/BC. For each particular problem, solutions were computed employing 50, 100, 200 and 400 boundary segments. The distribution of these segments accounted for the fact that in practice AB >> BC, and accordingly the sides AB and CD were sub-divided into four times as many segments as the sides BC and DA. In all cases the

Newton-Raphson process converged with fewer than 10 iterations. Furthermore, for a given set of the system parameters, the number of iterations taken by the Newton-Raphson procedure was independent of the number of boundary segments employed. In order to check the accuracy of the Newton-Raphson process solutions were obtained for problems with $N_c = 0$. These solutions were found to be identical to those given by the standard (linear) BIE implementation.

Representative results for three particular problems are presented in Tables 1, 2 and 3, which show the potential ϕ at the lattice points of a uniform rectangular mesh and also the respective heat transfer rate Q, energy balance E, and the number of iterations taken by the Newton-Raphson solution process. Also shown in Tables 1, 2 and 3 are the Richardson entrapolation [14] limits for the heat transfer rate Q, based upon the results obtained for the 100, 200 and 400 segment discretizations. These results illustrate various features of the non-linear BIE formulation and are characteristic of those observed for other values of the system parameters. The results in Tables 1, 2 and 3 correspond to the problems,

1. $Bi = 1$, $N_c = 0$ and $AB/BC = 10$,

2. $Bi = 0$, $N_c = 1$ and $AB/BC = 10$,

3. $Bi = 1$, $N_c = 1$ and $AB/BC = 10$,

respectively, and essentially describe the performance of a particular fin under different operating conditions. Problem 1 represents a situation in which convective heat transfer is dominant, such as in air-cooled industrial heat exchangers, while problem 2 corresponds to the case where heat dissipation is purely radiative, as for example, in space applications. Problem 3 relates to the situation where both convection and radiation are evident and, in this case, equally weighted. For the linear problem (1) the Newton-Raphson process required 2 iterations to determine the solution; the second iteration was necessary in order to satisfy the convergence criteria that the relative change in each iterate be less than 10^{-6} for successive iterations. For problems involving non-linear terms, e.g. 2 and 3, the number of iterations taken by the Newton-Raphson solution process depends in particular on the magnitude of the non-linearity. The radiation coefficient is the same in problems 2 and 3, and it is therefore expected that the Newton-Raphson method will take approximately the same number of iterations in these two problems. In fact, in both problems 2 and 3 the Newton-Raphson solution process converges with only 6 iterations, irrespective of the boundary discretization. In all three cases considered the temperature distribution ϕ and the heat transfer rate Q both display convergent behaviour as the boundary discretization is refined. Furthermore, the accuracy with which the energy conservation requirement $E = 0$ is satisfied improves as the number of boundary segments is increased. In fact closer inspection of the results displayed in Tables 1, 2 and 3 reveals that for each particular value of N the corresponding value of E/Q is almost identical in all three problems. As only the fin-dimensions, (i.e. AB/BC), are the same in these three problems it appears that the accuracy of the BIE solution is dependent only on the boundary discretization. This has been confirmed by the results obtained for other values of the parameter AB/BC.

On the boundary DA and, in particular, at the corners A and D, where the potential is prescribed to be unity, the BIE solutions are somewhat inaccurate. However, this inaccuracy cannot be attributed to the non-linearity as it is also evident when the non-linear term is absent, e.g. problem 1. This inaccuracy is probably due to the crudeness of the piecewise-constant approximation and therefore should be reduced by the introduction of piecewise-linear or piecewise-quadratic boundary variation, e.g. [8,9,10].

The results presented in Tables 1, 2 and 3 also illustrate certain aspects of the physics of the non-linear fin problem. The fin temperature distribution for the case where heat dissipation is purely radiative, (problem 2), is considerably higher than for the problems involving convection heat transfer, (1 and 3). However, the heat transfer rate for problem 2 is between 40 and 50 per cent less than for problems 1 and 3. This indicates that radiation heat transfer is less efficient than convection. Furthermore, comparison of the solutions for problems 1 and 3 shows that the inclusion of radiation causes a slight reduction in the temperature distribution, but still facilitates an increase in the heat transfer.

A particularly significant feature of the BIE solutions is the two-dimensional variation of the temperature distribution, which suggests that the previously employed one-dimensional analyses may be inaccurate.

CONCLUSIONS

A BIE formulation has been presented which enables the solution of plane Laplacian problems involving non-linear boundary conditions. This technique is shown to provide an effective treatment of a physical problem involving highly non-linear boundary conditions. The suitability of the presented technique, to the problem investigated, is highlighted by the considerable rapidity with which the iterative solution process converges. However, it must be emphasised that the non-linear BIE formulation is equally applicable to other two-dimensional potential problems involving non-linear boundary conditions, e.g. problems in water-wave theory.

REFERENCES

1. M.A. Jaswon and J.T. Symm, Integral Equation Methods in Potential Theory and Electrostatics, Academic Press, London, 1977.

2. G.D. Smith, Numerical Solution of Partial Differential Equations, Oxford University Press, 1972.

3. O.C. Zienkiewicz, The Finite Element Method in Engineering, McGraw-Hill, London, 1971.

4. F.J. Rizzo and D.J. Shippy, "A Method of Solution for Certain Problems of Transient Heat Conduction", AIAA Journal, Vol. 8, pp. 2004-2009, 1970.

5. G.T. Symm, "Treatment of Singularities in the Solution of Laplace's Equation by an Integral Equation Method", National Physical Laboratory, Report NAC 31, 1973.

6. M. Maiti and S.K. Chakrabarty, "Integral Equation Solutions for Simply Supported Polygonal Plates", International Journal of Engineering Science, Vol. 12, pp. 793-806, 1974.

7. W.A. Bell, W.L. Meyer and B.T. Zinn, "Predicting the Acoustics of Arbitrarily Shaped Bodies Using an Integral Approach", AIAA Journal, Vol. 15, pp. 813-820, 1977.

8. Y.S. Wu, F.J. Rizzo, D.J. Shippy and J.A. Wagner, "An Advanced Boundary Integral Equation Method for Two-Dimensional Electromagnetic Field Problems", Electric Machines and Electro-mechanics", Vol. 1, pp. 301-313, 1977.

9. G. Fairweather, F.J. Rizzo, D.J. Shippy and Y.S. Wu, "On the Numerical Solution of Two-Dimensional Potential Problems by an Improved Boundary Integral Equation Method", Journal of Computational Physics, Vol. 31, pp. 96-112, 1979.

10. D.B. Ingham, P.J. Heggs and M. Manzoor, "The Numerical Solution of Plane Potential Problems by Improved Boundary Integral Equation Methods", Submitted to the Journal of Computational Physics, 1980.

11. G.T. Symm, "Integral Equation Methods in Potential Theory II". Proceedings of the Royal Society, A275, pp. 33-46, 1963.

12. R. Butterfield, "The Application of the Integral Equation Methods to Continuum Problems in Soil Mechanics", Roscoe Memorial Symposium, Cambridge, 1972.

13. P.K. Banerjee, "Non-Linear Problems of Potential Flow", In Developments in Boundary Element Methods, Vol. 1, Edited by P.K. Banerjee and R. Butterfield, Applied Science Publishers, London, 1979.

14. A. Ralston, A First Course in Numerical Analysis, McGraw-Hill, New York, 1965.

15. E.M. Sparrow and R.D. Cess, Radiation Heat Transfer, Brooks/Cole, Belmont, 1970.

16. D.Q. Kern and A.D. Kraus, Extend Surface Heat Transfer, McGraw-Hill, New York, 1972.

17. R.C. Donovan and W.M. Rohrer, "Radiative and Convective Conducting Fins on a Plane Wall Including Mutual Irradiation", Journal of Heat Transfer, Vol. 93, pp. 41-46, 1971.

18. M.N. Schnurr, "Radiation From an Array of Longitudinal Fins of Triangular Profile", AIAA Journal, Vol. 13, pp. 691-693, 1975.

19. R.D. Karam and R.J. Eby, "Linearised Solution of Conducting Radiating Fins", AIAA Journal, Vol. 16, pp 536-538, 1978.

Table 1 : BIE Solution for the Problem 1

1.0589	0.5682	0.3095	0.1706	0.0994	0.0682
1.0309	0.5545	0.3008	0.1661	0.0977	0.0683
1.0157	0.5495	0.2977	0.1645	0.0971	0.0684
1.0079	0.5477	0.2966	0.1640	0.0969	0.0685
1.0030	0.5687	0.3100	0.1709	0.0996	0.0698
1.0000	0.5537	0.3005	0.1659	0.0975	0.0687
1.0000	0.5484	0.2972	0.1642	0.0969	0.0684
1.0000	0.5466	0.2961	0.1637	0.0967	0.0684
1.0000	0.5648	0.3079	0.1698	0.0990	0.0691
1.0000	0.5504	0.2987	0.1649	0.0970	0.0682
1.0000	0.5453	0.2955	0.1633	0.0963	0.0680
1.0000	0.5435	0.2944	0.1627	0.0961	0.0680
1.0000	0.5594	0.3049	0.1681	0.0980	0.0684
1.0000	0.5451	0.2958	0.1633	0.0960	0.0675
1.0000	0.5400	0.2926	0.1617	0.0954	0.0673
1.0000	0.5382	0.2915	0.1612	0.0952	0.0673
1.0047	0.5524	0.3010	0.1660	0.0967	0.0677
1.0000	0.5378	0.2918	0.1611	0.0947	0.0667
1.0000	0.5326	0.2886	0.1595	0.0941	0.0664
1.0000	0.5309	0.2875	0.1589	0.0939	0.0664
1.0636	0.5409	0.2946	0.1624	0.0946	0.0647
1.0374	0.5280	0.2864	0.1582	0.0930	0.0650
1.0213	0.5232	0.2834	0.1566	0.0924	0.0651
1.0119	0.5215	0.2824	0.1561	0.0922	0.0652

N = 50
    100
    200
    400

N

	50	100	200	400	limit
Q	0.3206	0.3147	0.3124	0.3116	0.3111
E/Q	0.0449	0.0166	0.0059	0.0020	0.0000
Iterations	2	2	2	2	*

Table 2 : BIE Solution for the Problem 2

1.0346	0.7564	0.6180	0.5386	0.4932	0.4716
1.0181	0.7496	0.6136	0.5359	0.4918	0.4712
1.0092	0.7471	0.6119	0.5349	0.4913	0.4711
1.0046	0.7461	0.6113	0.5346	0.4911	0.4711
1.0018	0.7567	0.6183	0.5387	0.4933	0.4727
1.0000	0.7492	0.6134	0.5358	0.4917	0.4714
1.0000	0.7465	0.6117	0.5348	0.4912	0.4711
1.0000	0.7455	0.6111	0.5344	0.4910	0.4710
1.0000	0.7544	0.6172	0.5382	0.4929	0.4722
1.0000	0.7473	0.6125	0.5353	0.4913	0.4711
1.0000	0.7447	0.6108	0.5343	0.4908	0.4708
1.0000	0.7437	0.6102	0.5340	0.4907	0.4707
1.0000	0.7513	0.6158	0.5373	0.4923	0.4717
1.0000	0.7443	0.6112	0.5345	0.4908	0.4706
1.0000	0.7417	0.6095	0.5335	0.4903	0.4703
1.0000	0.7408	0.6089	0.5332	0.4901	0.4702
1.0031	0.7472	0.6140	0.5362	0.4915	0.4712
1.0000	0.7401	0.6093	0.5334	0.4900	0.4700
1.0000	0.7376	0.6076	0.5324	0.4895	0.4696
1.0000	0.7367	0.6070	0.5320	0.4893	0.4695
1.0381	0.7407	0.6109	0.5344	0.4903	0.4691
1.0233	0.7347	0.6067	0.5319	0.4889	0.4688
1.0139	0.7324	0.6051	0.5309	0.4884	0.4687
1.0081	0.7315	0.6046	0.5306	0.4883	0.4686

N = 50
     100
     200
     400

N

	50	100	200	400	limit
Q	0.1973	0.1944	0.1930	0.1924	0.1919
E/Q	0.0439	0.0166	0.0061	0.0021	0.0000
Iterations	6	6	6	6	*

Table 3 : BIE Solution for the Problem 3

1.0649	0.5444	0.2937	0.1615	0.0937	0.0639
1.0340	0.5320	0.2861	0.1577	0.0925	0.0646
1.0173	0.5273	0.2833	0.1563	0.0921	0.0649
1.0087	0.5256	0.2822	0.1558	0.0920	0.0650
1.0033	0.5451	0.2942	0.1618	0.0940	0.0655
1.0000	0.5312	0.2857	0.1575	0.0924	0.0650
1.0000	0.5262	0.2827	0.1560	0.0920	0.0649
1.0000	0.5245	0.2817	0.1555	0.0918	0.0649
1.0001	0.5407	0.2921	0.1607	0.0933	0.0648
1.0000	0.5275	0.2840	0.1566	0.0919	0.0645
1.0000	0.5227	0.2811	0.1551	0.0914	0.0645
1.0000	0.5210	0.2800	0.1546	0.0913	0.0645
1.0000	0.5346	0.2892	0.1591	0.0924	0.0642
1.0000	0.5216	0.2811	0.1550	0.0910	0.0639
1.0000	0.5169	0.2783	0.1536	0.0905	0.0639
1.0000	0.5152	0.2772	0.1531	0.0904	0.0639
1.0059	0.5268	0.2854	0.1571	0.0912	0.0635
1.0000	0.5137	0.2773	0.1529	0.0898	0.0631
1.0000	0.5089	0.2744	0.1515	0.0893	0.0630
1.0000	0.5073	0.2734	0.1510	0.0891	0.0630
1.0715	0.5141	0.2792	0.1537	0.0892	0.0607
1.0439	0.5031	0.2721	0.1501	0.0881	0.0614
1.0262	0.4987	0.2694	0.1488	0.0877	0.0617
1.0154	0.4972	0.2684	0.1483	0.0876	0.0619

N = 50
    100
    200
    400

	N				
	50	100	200	400	limit
Q	0.3712	0.3664	0.3639	0.3627	0.3617
E/Q	0.0443	0.0168	0.0061	0.0022	0.0000
Iterations	6	6	6	6	*

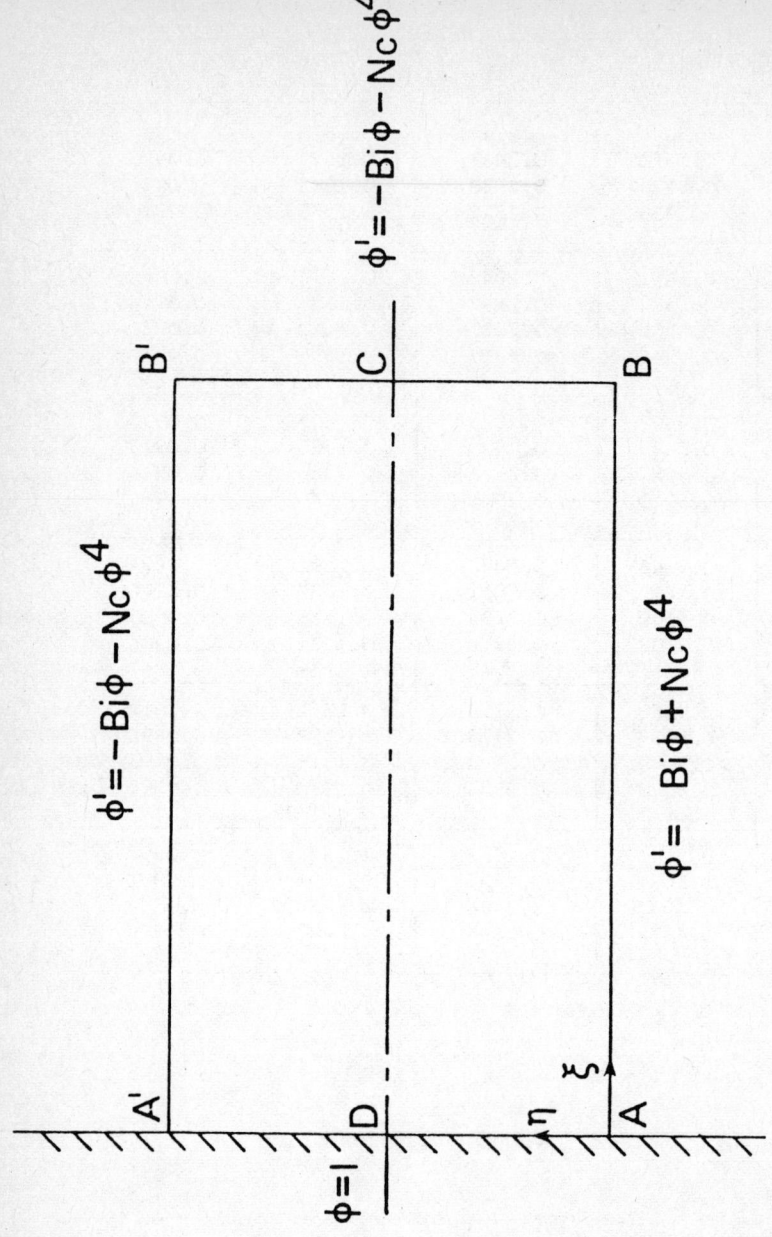

Fig.1 Schematic representation of an isolated fin

4.2 IMPROVED FORMULATIONS FOR THE ANALYSIS OF CONVECTING AND RADIATING FINNED SURFACES

ABSTRACT

The study of the heat flow within finned surfaces involving combined convective and radiative heat dissipation is conventionally based upon uni-directional analyses with attention restricted solely to the fin side. In this study the heat conduction within the interface to which the fins are attached and the heat transfer from the unfinned side of this interface are also considered. The general situation in which the fins and base-surface have different thermal conductivities and different surface emissivities is examined. Fin-to-base-surface, fin-to-environment, fin-to-fin and base-surface-to-environment radiant interactions are all accounted for, but the convective fluids are assumed to be radiatively transparent. One-and two-dimensional formulations are developed for analysing the heat flow. These represent a major extension of the previously employed formulations. The results obtained indicate, in particular, that the previously employed formulations are inadequate for the effective design of finned heat exchangers.

INTRODUCTION

The theoretical study of the heat flow within finned heat exchangers is of considerable practical importance because of the extensive utilization of finned surfaces for heat transfer enhancement in applications varying from air-cooled heat exchangers in the process industries to heat rejection equipment in space vehicles. The accurate prediction of the thermal performance of finned heat exchangers is essential for compact and efficient design. However, the analysis of such devices is conventionally based upon several simplifying assumptions, in particular, that the heat flow is uni-directional. This assumption facilitates a considerable reduction in the complexity of the analysis. In fact, for problems involving purely convective heat dissipation, the one-dimensional approximation, in general, permits an analytical treatment, e.g. [1,2,3].

The applicability of the one-dimensional approximation has been extensively investigated, but only for problems involving purely convective surface heat transfer, [4-9]. The early investigations restricted attention solely to the fin side and concluded that two-dimensional effects are negligible provided the transverse Biot number is much less than unity, [4,5,6]. However, recent investigations of the combined fin and supporting surface have shown that the presence of fins induces two-dimensional effects within the supporting surface and these may in turn act to produce two-dimensional variations within the fin, [7,8,9]. Suryanarayana [8] has reported that the difference between fin heat transfer rates predicted by one- and two-dimensional analyses can be as much as 80 per cent. It is therefore essential for the effective design of finned heat exchangers to consider the complete fin assembly and to employ a multi-dimensional analysis.

The study of the heat flow within finned surfaces involving radiative heat dissipation is particularly relevant to the design of heat exchangers operating at high temperatures or in atmosphere-free environments, [10-15]. The analysis of such systems has progressed considerably since the original investigations which examined isolated black fins operating in the absence of any surface convection, e.g.

[10,11]. The more recent investigations have considered combined convective and radiative heat dissipation from arrays of gray fins which radiatively interact with adjacent fins and also the base-surface, e.g. [14,15]. However, even these advanced formulations are unsuitable for the effective design of finned heat exchangers because they are based upon uni-directional heat flow analyses with attention restricted solely to the fin side, i.e. they fail to account for the thermal interaction between the fins and the supporting surface.

The primary objective of the present investigation is to develop an accurate formulation for analysing the heat flow within finned surfaces which dissipate heat by both convection and radiation. First, a one-dimensional formulation is devised which considers the complete fin assembly, i.e. both the fins and the supporting surface simultaneously. The general situation in which the fins and the supporting surface have different thermal conductivities and different emissivities is examined. This formulation in itself represents a major extension of the previous work, e.g. [10-15], as the previous studies have restricted attention solely to the fin side and have only considered situations in which the fins and the base-surface have the same emissivity. However, a further and more fundamental extension is also devised. The general fin assembly situation is analysed on the basis of two-dimensional heat flow. This represents a considerable increase in the complexity of the analysis. However, it is shown that a recently developed implementation of the boundary integral equation (BIE) method [16] is capable of handling these complexities without difficulty, and gives accurate solutions with relatively modest demands of computational storage and time.

ANALYSIS

Assumptions

Consider a heat exchanger comprised of equally spaced longitudinal rectangular fins attached to a plane wall, as depicted schematically in Fig. 1. In this study a theoretical representation of this device is developed on the basis of the following assumptions:

1. the wall and fin materials are isotropic and have constant thermal conductivities

2. there is perfect contact between the wall and the fins

3. convective heat exchange is governed by Newton's Law of Cooling

4. the surface heat exchange on the unfinned side of the fin assembly is purely convective, whilst that on the finned side is a combination of convection and radiation

5. all radiating surfaces are opaque, grey and diffuse

6. the radiative environment behaves as an isothermal black surface

7. the convective fluids are radiatively transparent.

Assumptions (1), (2), (3), (5), (6) and (7) are consistent with those conventionally employed in the analysis of convecting and radiating finned surfaces, e.g. [14,15]. Assumption (4) indicates that the radiative heat exchange on the unfinned side of the fin assembly is to be neglected. This simplification is introduced primarily in order to facilitate a reduction in the complexity of the analysis, but it is not unrepresentative of the actual physical situation. In practical applications of finned heat exchangers the unfinned side convection is relatively high and consequently only at very elevated temperatures is there any appreciable radiative contribution to the total heat transfer, [12-15].

One-Dimensional Formulation

For steady-state one-dimensional heat conduction, the determination of the fin assembly temperature distribution requires the simultaneous solution of the energy equations,

$$\frac{d^2}{dx^2} \phi_w(X) = 0, \qquad \text{within the wall} \qquad (1)$$

and

$$\frac{d^2}{dx^2} \phi_f(X) - \frac{Bi_2}{KT} (\phi_f(X) - \phi_{2f}) - \varepsilon_f \frac{Nc_2}{KT} (\phi_f^4(X) - G_f(X)) = 0,$$
$$\text{within the fin} \qquad (2)$$

combined with the irradiation equations,

$$G_w(0) = \int_{X=0}^{L} F_{FG-dX} (\varepsilon_f \phi_f^4(X) + (1-\varepsilon_f) G_f(X))$$

$$+ \int_{X'=0}^{L} F_{FG-dX'} (\varepsilon_f \phi_f^4(X) + (1-\varepsilon_f) G_f(X))$$

$$+ F_{FG-EE} \cdot \phi_{2e}^4 \qquad (3)$$

and

$$G_f(X) = F_{dX-FF'} (\varepsilon_w \phi_w^4(0) + (1-\varepsilon_w) G_w(0))$$

$$+ \int_{X'=0}^{L} F_{dX-dX'} (\varepsilon_f \phi_f^4(X) + (1-\varepsilon_f) G_f(X))$$

$$+ F_{dX-EE'} \phi_{2e}^4 \tag{4}$$

subject to the boundary conditions,

$$X=-W: \quad \frac{d}{dX}\phi_w(X) = -Bi_1(1-\phi_w(X)) \tag{5a}$$

$$X=0: \quad \phi_w(X) = \phi_f(X) \tag{5b}$$

$$X=0: \quad \frac{d}{dX}\phi_w(X) = KT\frac{d}{dX}\phi_f(X) - Bi_2(1-T)(\phi_w(0)-\phi_{2f}) - \varepsilon_w Nc_2(1-T)(\phi_w^4(0)-G_w(0)) \tag{5c}$$

$$X=L: \quad \frac{d}{dX}\phi_f(X) = -\frac{Bi_2}{K}(\phi_f(X)-\phi_{2f}) - \frac{Nc_2}{K}(\phi_f^4(X)-\phi_{2e}^4) \tag{5d}$$

The energy equation (1) is obtained by performing an energy balance on an infinitesimal element of the wall, whilst the energy equation (2) and the irradiation equations (3) and (4) are obtained by performing an energy balance on an infinitesimal element in an arbitrary fin and combining this with the radiant flux balances between two-adjacent fins, the included base-surface and the environment, [14, 15].

The boundary condition (5a) relates to the convective heat exchange of the unfinned side of the fin assembly, whilst the boundary condition (5d) represents the combined convective and radiative heat dissipation from the fin-tip. The boundary conditions (5b) and (5c) arise from the assumption of perfect wall-to-fin contact which requires that the temperature and heat flux be continuous across the contact interface.

The mathematical complexity of the problem described by equations (1) - (5) precludes a completely analytical treatment. However, closer inspection of these equations indicates that a certain amount of analytical manipulation is possible. The differential equation (1) can be integrated exactly to give,

$$\phi_w(X) = \alpha X + \beta \qquad (6)$$

where α and β are, as yet, undetermined constants. This relation may be re-expressed in the form

$$\phi_w(X) = \frac{-\phi_a X + \phi_b (W+X)}{W} \qquad (7)$$

where ϕ_a and ϕ_b denote the unknown surface temperatures of the wall; $\phi_a = \phi_w(X=-W)$ and $\phi_b = \phi_w(X=0)$. Combining equation (7) with the boundary condition (5a) gives rise to a relation between ϕ_a and ϕ_b, and accordingly equation (7) becomes

$$\phi_w(X) = \frac{-Bi_1 X + \phi_b(Bi_1 X + (1+Bi_1 W))}{(1+Bi_1 W)} \qquad (8)$$

Thus, it only remains to determine the fin temperature distribution, ϕ_f. By virtue of the perfect contact condition (5b) the fin-base temperature is identically equal to ϕ_b, and consequently once the fin temperature distribution ϕ_f is determined then the wall temperature distribution ϕ_w can be simply computed from the relation (8).

In order to determine the fin temperature distribution a modified implementation of the finite-difference method [17] is devised. The modification is necessary in order to incorporate the radiation viewfactors which are explicitly defined only for the radiant interaction between isothermal surfaces, [18]. The fin length is subdivided into N-1 equal-sized elements of length H and nodes are situated at the endpoints of each of these elements, as shown schematically in Fig. 2. Assuming that the surface from $x_i - \frac{H}{2}$ to $x_i + \frac{H}{2}$ has constant temperature ϕ_i and uniform irradiation G_i, (with appropriate modifications for x_1 and x_N), and then discretizing equation (2) using central difference approximations for the derivatives gives,

$$\phi_{i-1} - (2+H^2 \frac{Bi_2}{KT}) \phi_i + \phi_{i+1} - H^2 \varepsilon_f \frac{Nc_2}{KT} (\phi_i^4 - G_i) = - H^2 \frac{Bi_2}{KT} \phi_{2f} \qquad (9)$$

Collocating the expression (9) at each of the nodal points $x_1, x_2, \ldots, x_N$ generates a system of N non-linear algebraic equations involving 2N+2 unknowns, namely, $\phi_0, \phi_1, \ldots, \phi_{N+1}$ and $G_1, G_2, \ldots, G_N$. An additional N equations are generated by substituting for $G_w(0)$ from equation (3) into equation (4) and then collocating the discretized form of equation (4) at each of the nodal points $x_1, x_2, \ldots, x_N$. Thus, the combined algebraic representation constitutes 2N equations in 2N+2 unknowns.

The number of unknowns is reduced to 2N by enforcing the boundary conditions (5c) and (5d). In boundary condition (5c), prior to discretization, all occurrences of $\phi_w(x)$ are appropriately replaced using the expression (8), and $G_w(0)$ is replaced using the relation (3). Then, discretizing, with central difference approximations for the derivative $\frac{d}{dX} \phi_f(X)$ gives rise to an explicit expression for ϕ_0 in terms of $\phi_1, \phi_2, \ldots, \phi_N$ and $G_1, G_2, \ldots, G_N$. Similarly, discretizing boundary condition (5d) gives an explicit expression for ϕ_{N+1} in terms of ϕ_{N-1}. Thus, ϕ_0 and ϕ_{N+1} can be eliminated from the algebraic representation leaving a system of 2N non-linear algebraic equations involving 2N unknowns, namely, $\phi_1, \phi_2, \ldots, \phi_N$ and $G_1, G_2, \ldots, G_N$. The solution to this system of non-linear algebraic equations is determined using the Newton-Raphson method [19] which is particularly appropriate for this problem as the differentiations for the elements of the Jacobian matrix [19] can be performed explicitly.

Two-Dimensional Formulation

For steady-state two-dimensional conductive heat flow, the inherent thermal symmetry indicates that it is only necessary to examine that section of the fin assembly bounded by the contour ABCDEFGA, Fig. 1. Thus, the determination of the fin assembly temperature distribution requires the simultaneous solution of

$$\left(\frac{\partial^2}{\partial X^2} + \frac{\partial^2}{\partial Y^2}\right)\phi_w(X,Y) = 0 \qquad \text{within the region ABCFGA, Fig. 3,} \tag{10}$$

and

$$\left(\frac{\partial^2}{\partial X^2} + \frac{\partial^2}{\partial Y^2}\right)\phi_f(X,Y) = 0 \qquad \text{within the region CDEFC, Fig. 3,} \tag{11}$$

subject to

on AB : $\dfrac{\partial}{\partial X}\phi_w(X,Y) = -\text{Bi}_1(1-\phi_w(X,Y))$ \hfill (12a)

on BC : $\dfrac{\partial}{\partial Y}\phi_w(X,Y) = 0$ \hfill (12b)

on CD : $\dfrac{\partial}{\partial Y}\phi_f(X,Y) = 0$ \hfill (12c)

on DE : $\dfrac{\partial}{\partial X}\phi_f(X,Y) = -\dfrac{\text{Bi}_2}{K}(\phi_f(X,Y) - \phi_{2f}) - \varepsilon_f\dfrac{\text{Nc}_2}{K}(\phi_f^{\,4}(X,Y) - \phi_{2e}^{\,4})$ \hfill (12d)

on EF : $\dfrac{\partial}{\partial Y}\phi_f(X,Y) = -\dfrac{\text{Bi}_2}{K}(\phi_f(X,Y) - \phi_{2f}) - \varepsilon_f\dfrac{\text{Nc}_2}{K}(\phi_f^{\,4}(X,Y) - G_f(X))$ \hfill (12e)

on FC : $\phi_w(X,Y) = \phi_f(X,Y)$ \hfill (12f)

and $\dfrac{\partial}{\partial X}\phi_w(X,Y) = K\dfrac{\partial}{\partial X}\phi_f(X,Y)$ \hfill (12g) on FG : $\dfrac{\partial}{\partial X} \phi_w(X,Y) = -\text{Bi}_2(\phi_w(X,Y) - \phi_{2f}) - \varepsilon_w \text{Nc}_2(\phi_w^4(X,Y) - G_w(Y))$

(12h)

on GA : $\dfrac{\partial}{\partial Y} \phi_w(X,Y) = 0$ (12i)

combined with the irradiation equations,

$$G_w(Y) = \int_{X=0}^{L} F_{dY-dX}(\varepsilon_f \phi_f^4(X,Y) + (1-\varepsilon_f) G_f(X))$$

$$+ \int_{X'=0}^{L} F_{dY-dX'}(\varepsilon_f \phi_f^4(X,Y) + (1-\varepsilon_f) G_f(X))$$

$$+ F_{dY-EE'} \phi_{2e}^4 \quad (13)$$

and

$$G_f(X) = \int_{Y=T}^{P} F_{dX-dY}(\varepsilon_w \phi_w^4(X,Y) + (1-\varepsilon_w) G_w(Y))$$

$$+ \int_{Y'=T}^{P} F_{dX-dY'}(\varepsilon_w \phi_w^4(X,Y) + (1-\varepsilon_w) G_w(Y))$$

$$+ F_{dX-EE'} \phi_{2e}^4 \quad (14)$$

The boundary conditions (12b), (12c) and (12i) arise from the geometrical symmetry of the fin arrangement and stipulate that there is no heat flux across the fictitious boundaries BC, CD and GA, Fig. 3. The boundary conditions (12f) and (12g) result from the assumption of perfect wall-to-fin contact which requires the temperature and heat flux to be continuous across the contact interface, FC, Fig. 3. The remaining boundary conditions describe the heat exchange from the exposed surfaces AB and DEFG.

The irradiation equations (13) and (14) are derived by considering the radiant flux within the enclosure EFF'E'E, Fig. 1, and include the effects of fin-to-environment, fin-to-fin, fin-to-wall and wall-to-environment radiant interactions. These equations are considerably more complicated than the corresponding relations in the one-dimensional formulation, namely, equations (3) and (4). This is due to the fact that with the two-dimensional formulation both the temperature and the surface irradiation vary along the base-surface FG, Fig. 1, whereas with the one-dimensional formulation the base surface behaves as an isothermal surface with uniform irradiation.

The solution to the problem described by equations (10), (11), (12), (13) and (14) is susceptible to treatment by various numerical techniques, e.g. the finite-difference [17], finite-element [20] and boundary integral equation (BIE) [21] methods. In this study a recently devised non-linear implementation of the BIE method [16] is employed because it offers several important advantages over the finite-difference and finite-element methods for the solution of this particular problem. The principal advantage is that, in contrast to the finite-difference and finite-element methods, the BIE discretization occurs only on the domain boundary and therefore generates a considerably smaller algebraic representation than an equivalent finite-difference or finite-element approximation [16]. Thus, in comparison with the finite-difference and finite-element methods, the BIE formulation facilitates substantial reductions in the computational storage and time requirements. In addition, the irradiation relations (13) and (14) are directly compatible with the classical BIE formulation. The BIE discretization process involves the subdivision of the domain boundary into isothermal segments and is therefore ideally suited for the discretization of the relations (13) and (14) in which the radiation viewfactors are explicitly defined only for the radiant interaction between isothermal surfaces.

With the BIE formulation, the problem described by equations (10) - (14) is first transformed into an integral equation by the application of Green's Integral Formula [16,21] to the governing equations (10) and (11), and the boundary conditions (12). This integral equation involves two coupled contour integrals, one around ABCFGA and the other around CDEFC, Fig. 3. The coupling arises through the interface boundary conditions (12f) and (12g). In order to effect a solution to this integral equation the contours ABCFGA and CDEFC are subdivided into a total of N rectilinear segments and nodes are situated at the midpoint of each of these segments. The temperature, heat flux and, where appropriate, irradiation on each segment are approximated by piecewise-constant functions. Then, the discretized form of the integral equation is collocated at each of the boundary nodes. This generates a system of N non-linear algebraic equations involving the N unknown nodal temperatures and, in addition, N(EF) + N(FG) unknown values of the irradiation from the segments on the surfaces EF and FG, Fig. 3. An additional N(EF) + N(FG) non-linear algebraic equations are generated by discretizing the irradiation relations (13) and (14) in accordance with the overall discretization. The solutions to the combined system of N + N(EF) + N(FG) non-linear algebraic equations involving N + N(EF) + N(FG) unknowns is achieved employing the Newton-Raphson method combined with a Gaussian elimination [19] technique for solving the linearized equations. As with the one-dimensional formulation, the Newton-Raphson iterative scheme is particularly suitable for this problem because the differentiations for the elements of the appropriate Jacobian matrix can be performed explicitly.

Viewfactors

In both the one- and two-dimensional formulations, viewfactors are used to describe the fraction of radiant energy leaving one surface and impinging upon a second surface. The expressions defining these viewfactors are derived on the basis that the surfaces emit and reflect radiation diffusely, assumption (5) [18]. These expressions involve the integration of relatively complicated displacement functions over the two relevant surfaces [18]. However, if end effects are neglected, then the viewfactors can be evaluated explicitly employing a relatively simple formula, namely Hottels Crossed-String Identity [18]. As end effects are neglected in all other aspects of the one- and two-dimensional formulations, for consistency, Hottel's Crossed-String Identity is employed for the evaluation of the viewfactors arising in these formulations.

HEAT EXCHANGER PERFORMANCE

The heat flow through the fin assembly is most conveniently expressed in the form of an augmentation factor, Aug, defined as the ratio of the heat transfer rate of the finned assembly to that of the unfinned wall operating under the same conditions. In order to evaluate the augmentation factor it is first necessary to determine the heat transfer rate of the unfinned wall. This requires the solution of the energy equation

$$\frac{d^2}{dX^2} \phi^*_w (X) = 0 \qquad \text{within the wall} \qquad (15)$$

subject to the boundary conditions

$$X = -W: \quad \frac{d}{dX} \phi^*_w (X) = -Bi_1 (1-\phi^*_w(X)) \qquad (16a)$$

$$X = 0: \quad \frac{d}{dX} \phi^*_w (X) = -Bi_2 (\phi^*_w(X) - \phi_{2f}) - \varepsilon_w Nc_2 (\phi^{*4}_w(X) - \phi_{2e}^4) \qquad (16b)$$

where the energy equation (15) is obtained by performing an energy balance on an infinitesimal element of the wall, and the boundary conditions (16) describe the heat exchange from the surfaces of the wall.

It is not possible to derive a closed form solution for the problem defined by equations (15) and (16) because of the non-linearities occurring in the boundary condition (16b). However, the need to use numerical techniques, such as the finite-difference or finite-element methods, can be avoided by some simple analytical manipulation of the equations (15) and (16). Integrating equation (15) exactly and then enforcing boundary condition (16a) gives

$$\phi^*_w (X) = \frac{Bi_1 X + \phi^*_b (Bi_1 X + (1+Bi_1 W))}{(1+Bi_1 W)} \qquad (17)$$

where ϕ^*_b denotes the unknown temperature of the surface $X=0$. Combining the relation (17) with the boundary condition (16b) gives a non-linear equation involving ϕ^*_b,

$$(1+Bi_2 W - \frac{1}{1+Bi_1 W}) \phi^*_b + \varepsilon_w Nc_2 W \phi^{*4}_b = \frac{Bi_1 W}{1+Bi_1 W} + Bi_2 W \phi_{2f} + \varepsilon_w Nc_2 W \phi_{2e}^4 \qquad (18)$$

The solution to equation (18) can be achieved employing the Newton-Raphson method. This determines the surface temperature ϕ_b^* and therefore the temperature distribution within the wall $\phi_w^*(X)$, equation (17). The corresponding heat transfer rate is given by,

$$Q^* = k_w \frac{d}{dX} \phi_w^*(X = -W) \; \theta_1$$

$$= \frac{Bi_1}{(1+Bi_1 W)} (1-\phi_b^*) \, k_w \, \theta_1 \qquad (19)$$

and therefore the augmentation factors corresponding to the one- and two-dimensional solutions for the fin assembly temperature distribution can be evaluated using,

$$AUG1 = Q1/Q^*$$

$$= \frac{(1-\phi_b)}{(1-\phi_b^*)} \qquad (20)$$

and

$$AUG2 = Q2/Q^*$$

$$= \frac{-(1+Bi_1 W)}{(1-\phi_b^*)} \int_{AB} (1-\phi_w(X,Y)) \, dS \qquad (21)$$

where dS denotes an incremental element of AB, Fig. 3. In the context of the BIE solutions, the integration (21) can be easily performed employing a quadrature consistent with the overall discretization.

ACCURACY OF NUMERICAL CALCULATIONS

The finite-difference method employed in the one-dimensional formulation and the BIE method employed in the two-dimensional formulation both involve the introduction of approximations in order to determine the fin assembly temperature distribution. Consequently, the corresponding solutions invariably include errors. The errors in both the finite-difference solutions and the BIE solutions are related to the associated mesh sizes [17,21] and thereford should diminish as the respective discretization is refined. In order to check for this convergence, for each particular problem considered, solutions are computed for three different levels of discretization corresponding to a uniform refinement of the appropriate mesh in the ratio of $1 : \frac{1}{2} : \frac{1}{4}$. The one-dimensional finite-difference solutions are computed by subdividing the fin length into 25, 50 and 100 elements, whilst the two-dimensional BIE solutions are obtained by subdividing the boundary contours ABCFGA and CDEFC, Fig. 3, into a total of 80, 160 and 320 segments. The distribution of the BIE boundary segments takes account of the fact that, in practice, the fin length is the largest dimension of the fin assembly action ABCDEFGA, Fig. 3, and consequently, the sides CD and EF are subdivided into four times as many segments as AX, XB, BC, CF, FG, GA, DE and FC, Fig. 3. This particular distribution of the boundary segments has been found to offer the most efficient use of the computational resources with respect to the accuracy of the corresponding solutions.

Solutions have been computed for a wide range of the system parameters Bi_1, Bi_2, K, L, T, W, Nc_2, ε_w, ε_f, ϕ_{2e} and ϕ_{2f}. It has been found that both the one-dimensional and the two-dimensional solutions display a monotonic convergent behaviour as the respective discretization is refined. The one-dimensional solutions converge quite rapidly;

the 50 and 100 element solutions for the augmentation factor AUG1 always agree to at least 4 significant figures. In contrast, the 160 and 320 segment solutions for AUG2 only agree to 2 significant figures. As further refinement of the boundary discretization is impractical, the limiting values of AUG2 are computed by extrapolation of the 80, 160 and 320 boundary segment solutions using Richardson's formula [19],

$$ERR(N) \quad \alpha \quad (H(N))^{\alpha} \qquad (22)$$

where ERR(N) is the error in the N segment solution, H(N) is an associated segment length and α is the order of the extrapolation. In order to determine the accuracy of the extrapolation, for each level of discretization, the augmentation factor was also evaluated by integrating the heat flux from the finned side of the fin assembly using,

$$AUGF2' = Bi_1 \frac{(1+Bi_1 W)}{(1-\phi_b^*)} \left\{ \int_{DE} (Bi_2(\phi_f(X,Y)-\phi_{2f}) + Nc_2(\phi_f^4(X,Y)-\phi_{2e}^4))dS \right.$$
$$+ \int_{EF} (Bi_2(\phi_f(X,Y)-\phi_{2f}) + Nc_2(\phi_f^4(X,Y)-G_f(X)))dS$$
$$\left. + \int_{FG} (Bi_2(\phi_w(X,Y)-\phi_{2f}) + Nc_2(\phi_w^4(X,Y)-G_w(Y)))dS \right\} \qquad (23)$$

and then these values were also extrapolated using Richardson's method, equation (22). It was found that the extrapolated values of AUG2 and AUG2' always agree to at least 3 significant figures.

The results presented in the following section are those predicted by the 100 element implementation of the finite-difference method, and those obtained by extrapolation of the 80, 160 and 320 segment BIE solutions.

RESULTS AND DISCUSSION

For both one- and two-dimensional formulations a complete description of the fin assembly heat transfer requires the specification of the eleven independent parameters Bi_1, Bi_2, K, L, T, W, Nc_2, ε_w, ε_f, ϕ_{2e} and ϕ_{2f}. The excessive computational time required to determine solutions for a complete range of values of the system parameters precludes the possibility of performing a complete parametric study. Instead, a practically more feasible procedure is adopted in order to investigate the effects of variations in the system parameters on the heat transfer capabilities of finned heat exchangers. For each particular problem the parameters K, L, T, W, ε_w and ε_f are assigned values and then solutions are computed for a comprehensive range of parameters Bi_1, Bi_2, Nc_2, ϕ_{2e} and ϕ_{2f}. This is essentially equivalent to investigating the performance of a given heat exchanger under different operating conditions. However, due to the limited available space for presentation of these results, only the solutions to selected problems are presented here.

Errors in the Conventional Analyses

The analysis of the heat flow within convecting and radiating finned surfaces is conventionally based upon the assumption that the temperature of the base-surface is unaffected by the addition of fins, e.g. [10-15]. An investigation has been performed in order to determine the applicability of this assumption. For each particular problem considered solutions were computed for both the unfinned wall and the fin assembly. The results for two particular problems are presented in Figs. 4 and 5 which show the temperature distributions within the unfinned wall and the fin assembly for the problems,

	Bi_1	Bi_2	K	L	T	W	Nc_2	ε_w	ε_f	ϕ_{2e}	ϕ_{2f}
A:	2×10^{-1}	10^{-3}	10	20	0.2	0.4	10^{-3}	0.6	0.8	0.1	0.2
B:	4.0	2×10^{-1}	20	20	0.2	0.4	10^{-3}	0.6	0.9	0.1	0.2

respectively. Problem A is representative of a heat exchanger comprised of a stainless-steel wall with aluminium fins operating with forced convection of air on the plain side and combined free-convection of air and radiation on the fin side. Problem B describes the performance of a heat exchanger consisting of a stainless-steel wall with copper fins, subject to forced-convection of water on the plain side and combined forced convection of air and radiation on the fin side.

The results presented in Figs. 4 and 5 are characteristics of those observed for other values of the system parameters and illustrate, in particular, that the addition of fins acts to depress the temperature of the base-surface. The reduction in the base-surface temperature, caused by the presence of the fins, is 7 per cent for problem A and over 25 per cent for problem B. In fact it has been found that for realistic values of the system parameters the reduction in the base-surface temperature can be as much as 50 per cent. Thus, the assumption that the base-surface temperature is unaffected by the addition of fins is clearly inappropriate.

In order to determine the extent of the inaccuracies in the heat transfer rates predicted by the isothermal base-surface analyses, solutions have been computed assuming that the base-surface temperature remains as predicted by the unfinned wall analysis. These solutions were obtained by a relatively simple algebraic manipulation of the fin assembly solutions. The temperature distribution within the fin, for the isothermal base-surface situation, is given by the relation,

$$\tilde{\phi}_f(X) = \phi_f(X) \frac{\phi_b^*}{\phi_b}$$

where $\phi_f(X)$ is the fin temperature distribution predicted by the fin assembly formulation, and ϕ_b^* and ϕ_b are the temperatures of the base-surface in the unfinned and finned states, respectively. The augmentation factor corresponding to this temperature distribution is given by,

$$AUG1 = Bi_1 \frac{(1+Bi_1)}{(1-\phi_b^*)} \quad Bi_2(1-T)(\tilde{\phi}_f(0) - \phi_{2f}) + \varepsilon_w Nc_2(1-T)(\tilde{\phi}_f^4(0) - G_w(0))$$

$$+ \int_{AB} (Bi_2(\tilde{\phi}_f(X) - \phi_{2f}) + \varepsilon_f Nc_2(\tilde{\phi}_f^4(X) - G_f(X))) dS$$

$$+ \quad Bi_2 T(\tilde{\phi}_f(L) - \phi_{2f}) + \varepsilon_f Nc_2 T(\tilde{\phi}_f^4(X) - \phi_{2e}^4)$$

where dS denotes an incremental element of AB.

A comparison of the augmentation factors AUG1 and $\widetilde{AUG1}$ has been performed for an extensive range of the system parameters. The results for problems A and B are displayed in Table 1. These results show that the errors in $\widetilde{AUG1}$ are in excess of 35 per cent for problem A and 90 per cent for problem B. In fact, it has been found that for some realistic problems the isothermal base-surface analysis can overpredict the heat exchanger performance by over 150 per cent. It is therefore apparent that the isothermal base-surface analysis is inadequate for the effective design of finned heat exchangers.

Errors in the One-Dimensional Analysis

As even the most advanced of the previously employed analyses for examining convecting and radiating finned surfaces have been based upon the assumption that the temperature of the base-surface is unaffected by the addition of fins, it may be concluded, from the predeeding discussion, that the one-dimensional fin assembly formulation presented in this study will facilitate an improvement in the design of finned heat exchangers. However, even this formulation may involve appreciable errors as it is based upon a uni-directional heat flow analysis; recent investigations [7,8,9], considering finned surfaces involving purely convective surface heat transfer, have shown that the one-dimensional approximation can result in errors of up to 80 per cent in the prediction of the fin heat transfer rate. Therefore, an investigation has been performed in order to determine the validity of the one-dimensional approximation for problems involving combined convective and radiative heat dissipation. Solutions have been computed for an extensive range of the system parameters employing both the one- and two-dimensional formulations presented in this study. The solutions to the problems A and B are displayed in Figs. 6 and 7. These solutions illustrate various deficiencies of the one-dimensional analysis and are representative of the solutions obtained for other values of the system parameters. In particular, in all the problems considered, it has been found that the two-dimensional solutions exhibit a temperature depression at the fin base, e.g. see Figs. 6 and 7. This temperature depression has been observed in the previous investigations considering two-dimensional heat flow within fin systems [7,8,9] and is attributed to the fact that a greater proportion of the heat flow is channelled through the fin-base than through the adjacent unfinned portion of the base-surface. However, this temperature depression is not accounted for in the one-dimensional analysis which takes the fin-base temperature to be identical to that at the adjacent unfinned portion of the base-surface. This deficiency of the one-dimensional formulation has been found to result in an under-estimation of the overall heat transfer by 2 per cent in problem A, and an over-estimation of the overall heat transfer rate by 27 per cent in problem B. However, errors or over 30 per cent have been observed in the heat transfer rates predicted by the one-dimensional formulation for problems with $Bi_1 < 10^{-1}$ and $Bi_2 < 10^{-2}$. Thus, it is apparent that a multi-directional analysis is essential for the accurate prediction of the performance of finned heat exchangers.

Errors in the Linear Analyses

In the vast majority of the published literature on the subject of fin heat transfer, the radiative heat dissipation is completely neglected, e.g.[1-9]. This simplification precludes the non-linear terms and therefore results in a considerable reduction in the complexity of the analysis. The errors which may accrue as a consequence of this simplification have not previously been analysed, although several investigators (e.g. [10,13]) have stated that these errors may be quite substantial for finned surfaces operating at high temperatures or in atmosphere free environments. Therefore an investigation has been performed in order to establish the extent of these errors. The two-dimensional fin assembly formulation developed in this study was employed for the purposes of this investigation because, as explained earlier, the other formulations include additional errors. For each particular problem examined, solutions were computed firstly for the case of combined convective and radiative heat dissipation and then for the case of purely convective surface heat transfer (i.e. for the case $Nc_2 = 0$). A large number of problems were examined with a view to determining quantitative limits on the applicability of the linear analysis. However, the complexity of the radiative heat transfer process is such that the results obtained were only sufficient to provide a qualitative indication of these limits. In particular, it was found that the results predicted by the linear and non-linear formulations can differ by up to 100 per cent if the order of magnitude of the convection parameter Bi_2 is less than or equal to that of the radiation parameter Nc_2, e.g. as may occur under free-convection type operating conditions. This may be attributed to the fact that under these conditions the radiative contribution to the overall heat flow is comparable to the convective contribution and therefore cannot be neglected. In contrast, if the convection parameter Bi_2 is appreciably larger than Nc_2, e.g. under forced convection conditions, then the error arising from neglecting the radiative heat transfer is virtually negligible because the convective heat dissipation dominates.

In order to illustrate these features, the results for the example problems A and B are given in Table 2. In problem A the fin side operating conditions involve combined free-convection of air and radiation, and consequently the augmentation factor AUG2 differs by nearly 70 per cent from that obtained if the radiative heat dissipation is neglected.

However, in problem B the fin side operating conditions involve forced-convection of air and this dominates to such an extent that when the radiative heat transfer is neglected the resulting change in the augmentation factor is less than 1 per cent.

CONCLUSIONS

One- and two-dimensional formulations have been developed for analysing the heat flow within finned heat exchangers which operate under conditions involving both convective and radiative surface heat dissipation. These formulations take into account the heat conduction within the interface to which the fins are attached and also the operating conditions on the unfinned side of the interface. A most significant feature of the results predicted by these formulations is the reduction in the temperature of the surface to which the fins are attached. As all previous formulations for finned surfaces involving radiative heat transfer have been based upon the assumption that the temperature of the base-surface is unaffected by the addition of fins, it is apparent that the analyses presented in this study will facilitate an improvement in the design of finned heat exchangers.

A comparison of the solutions predicted by the one- and two-dimensional fin assembly formulations shows that errors of up to 30 per cent can occur with the one-dimensional formulation. This is a consequence of the fact that the temperature depression occurring at the fin-base cannot be accurately modelled using a one-dimensional representation. Thus, the effective design of finned heat exchangers necessitates a two-dimensional analysis which takes account of the thermal interaction between the fins and the supporting surface.

Finally, an investigation into the effects of neglecting the radiative heat dissipation has indicated that, for problems involving free-convection type operating conditions, this may lead to errors of up to 100 per cent in the prediction of the heat transfer rate.

NOMENCLATURE

AUG1	one-dimensional augmentation factor
AUG2	two-dimensional augmentation factor
Bi_1	$= h_1 P/k_w$, Biot number
Bi_2	$= h_2 P/k_w$, Biot number
F_{A-B}	viewfactor from surface A to surface B
g	irradiation, W/m^2
G	$= g/\sigma\theta_1^4$, dimensionless irradiation
h	heat transfer coefficient, W/mK
k	thermal conductivity, $W/m^2 K$
l	fin length, m
L	$= l/P$
N	number of nodal points
Nc_2	$= \sigma P \theta_1^3/k_w$, dimensionless radiation parameter
P	half fin-pitch, m
Q	heat transfer rate, W
t	half fin-base thickness, m
T	$= t/P$
w	wall thickness, m
W	$= w/P$
x	longitudinal displacement, m
X	$= x/P$
y	transverse displacement, m

Y $= y/P$

θ temperature distribution, K

θ_1 plain-side fluid temperature, K

θ_{2f}, θ_{2e} fin-side fluid and fin-side environment temperatures, K

ϕ $= \theta/\theta_1$, dimensionless temperature distribution

σ Stefan-Boltzmann constant, $W/m^2 K^4$

K $= k_f/k_w$

Subscripts

1   plain-side

2   fin-side

f   fin

w   wall

Superscript

*   unfinned wall

REFERENCES

1. K.A. Gardner, "Efficiency of extended surface", Transactions of the ASME, Vol. 67, pp. 621-631, 1945.

2. S. Guceri and C.J. Maday, "A least weight circular cooling fin", Journal of Engineering for Industry, Vol. 97, pp. 1190-1193, 1975.

3. I. Mikk, "Convective fin of minimum mass", International Journal of Heat and Mass Transfer, Vol. 23, pp. 707-711, 1980.

4. R.K. Irey, "Errors in one-dimensional fin solution", Journal of Heat Transfer, Vol. 90, pp. 175-176, 1968.

5. M. Levistsky, "The criterion for validity of the fin approximation", International Journal of Heat and Mass Transfer, Vol. 15, pp. 1960-1963, 1972.

6. W. Lau and C.W. Tan, "Errors in one-dimensional heat transfer analysis in straight and annular fins", Journal of Heat Transfer, Vol. 95, pp. 549-551, 1973.

7. E.M. Sparrow and L. Lee, "Effects of fin-base temperature depression in a multifin array", Journal of Heat Transfer, Vol. 97, pp. 463-465, 1975.

8. N.V. Suryanarayana, "Two-dimensional effects on heat transfer from an array of straight fins", Journal of Heat Transfer, Vol. 99, pp. 129-132, 1977.

9. P.J. Heggs and P.R. Stones, "The effects of dimensions on the heat flowrate through extended surfaces", Journal of Heat Transfer, Vol. 102, 180-182, 1980.

10. R.L. Chambers and E.V. Somers, "Radiation fin efficiency for one-dimensional heat flow in a circular fin", Journal of Heat Transfer, Vol. 81, pp. 327-329, 1959.

11. J.G. Bartas and W.H. Sellers, "Radiation fin effectiveness", Journal of Heat Transfer, Vol. 82, pp. 73-75, 1960.

12. M.N. Schnurr and C.A. Cothran, "Radiation from an array of gray circular fins of trapezoidal profile", AIAA Journal, Vol. 12, 1476-1480, 1974.

13. M.N. Schnurr, "Radiation from an array of longitudinal fins of triangular profile", AIAA Journal, Vol. 13, pp. 691-693, 1975.

14. R.C. Donovan and W.M. Rohrer, "Radiative and convecting conducting fins on plane wall including mutual irradiation", Journal of Heat Transfer, Vol. 93, pp. 41-46, 1971.

15. R.G. Eslinger and B.T.F. Chung, "Periodic heat transfer in radiating and convecting fins or fin arrays", AIAA Journal, Vol. 17, pp. 1134-1140, 1979.

16. D.B. Ingham, P.J. Heggs and M. Manzoor, "Boundary integral equation solution of nonlinear plane potential problems", accepted for publication in Institute of Mathematics and Its Applications Journal of Numerical Analysis, 1981.

17. G.D. Smith, Numerical Solution of Partial Differential Equations, Oxford University Press, 1972.

18. H.C. Hottel and A.F. Sarofim, Radiative Transfer, McGraw-Hill, New York, 1967.

19. A. Ralston, A First Course in Numerical Analysis, McGraw-Hill, New York, 1965.

20. O.C. Ziekiewicz, The Finite Element Method in Engineering, McGraw-Hill, London, 1971.

21. M.A. Jaswon and G.T. Symm, Integral Equation Methods in Potential Theory and Electrostatics, Academic Press, London, 1977.

Table 1: Comparison of the solutions predicted by the conventional and fin assembly formulations.

	$\widetilde{\text{AUG1}}$	AUG1	$\widetilde{\text{AUG1}}$/AUG1
Problem A	14.48	10.68	135 per cent
Problem B	6.81	3.53	192 per cent

Table 2: Comparison of the solutions predicted by the linear and non-linear analyses.

	AUG2 (Nc_2=0)	AUG2	AUG2 (Nc_2=0)/AUG2
Problem A	18.26	10.85	168 per cent
Problem B	2.73	2.71	101 per cent

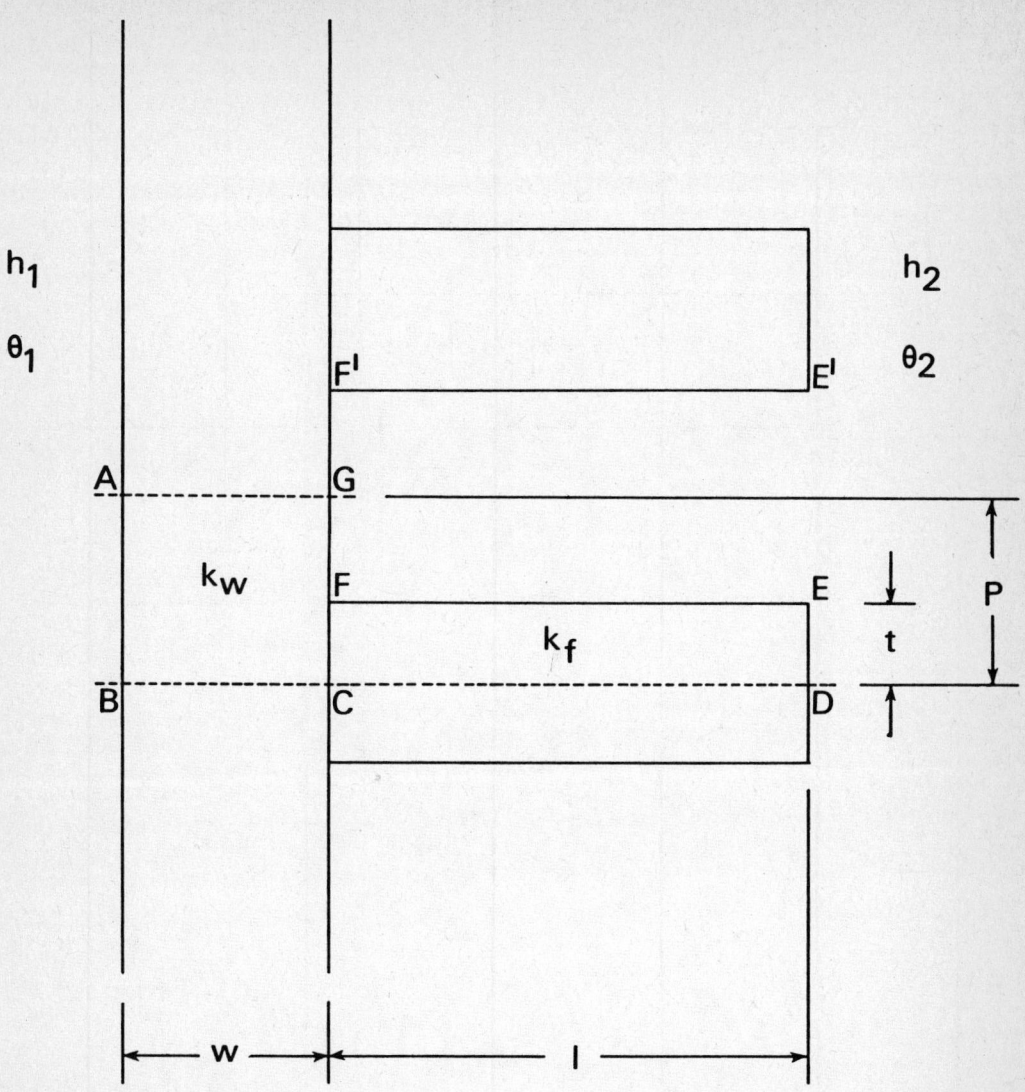

Fig.1 Schematic representation of the fin assembly

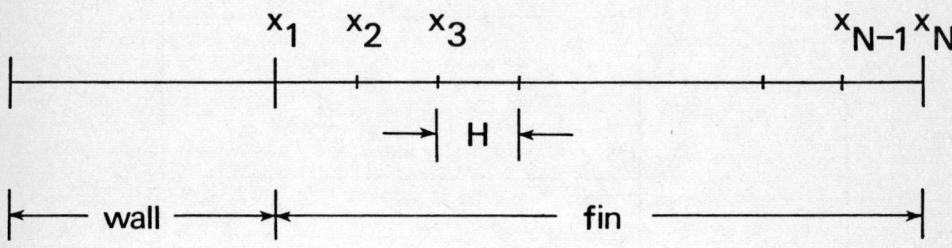

Fig.2 The finite-difference discretization for the one-dimensional model

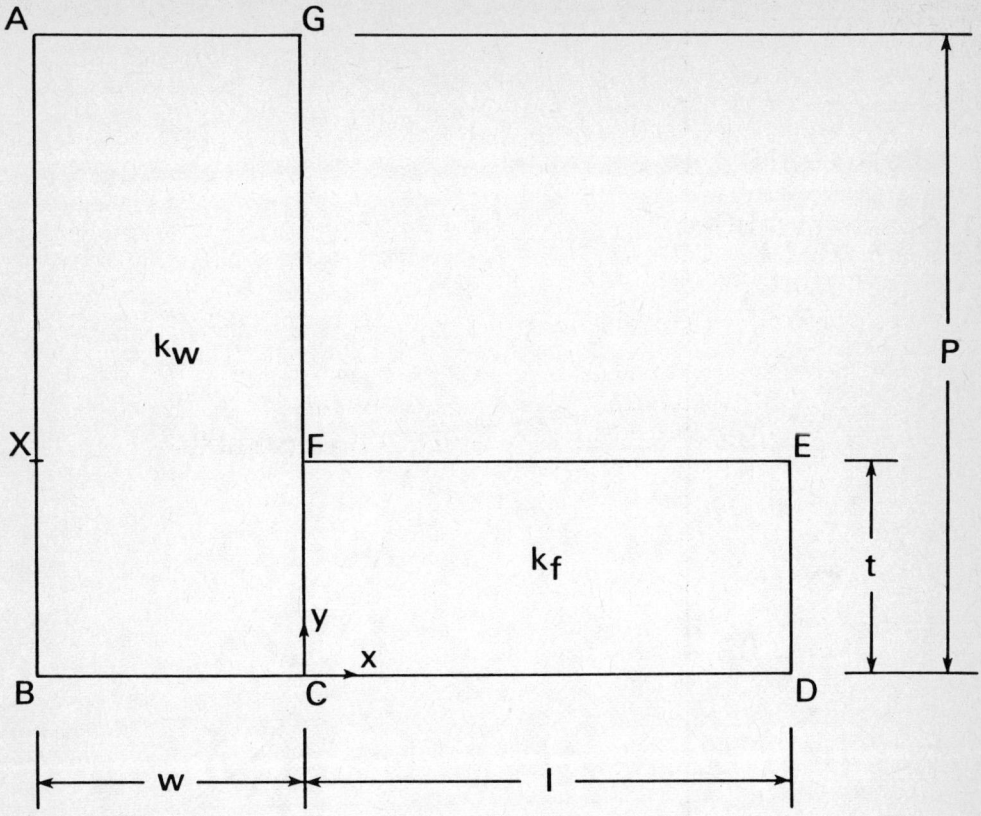

Fig.3 The region of geometrical and thermal symmetry

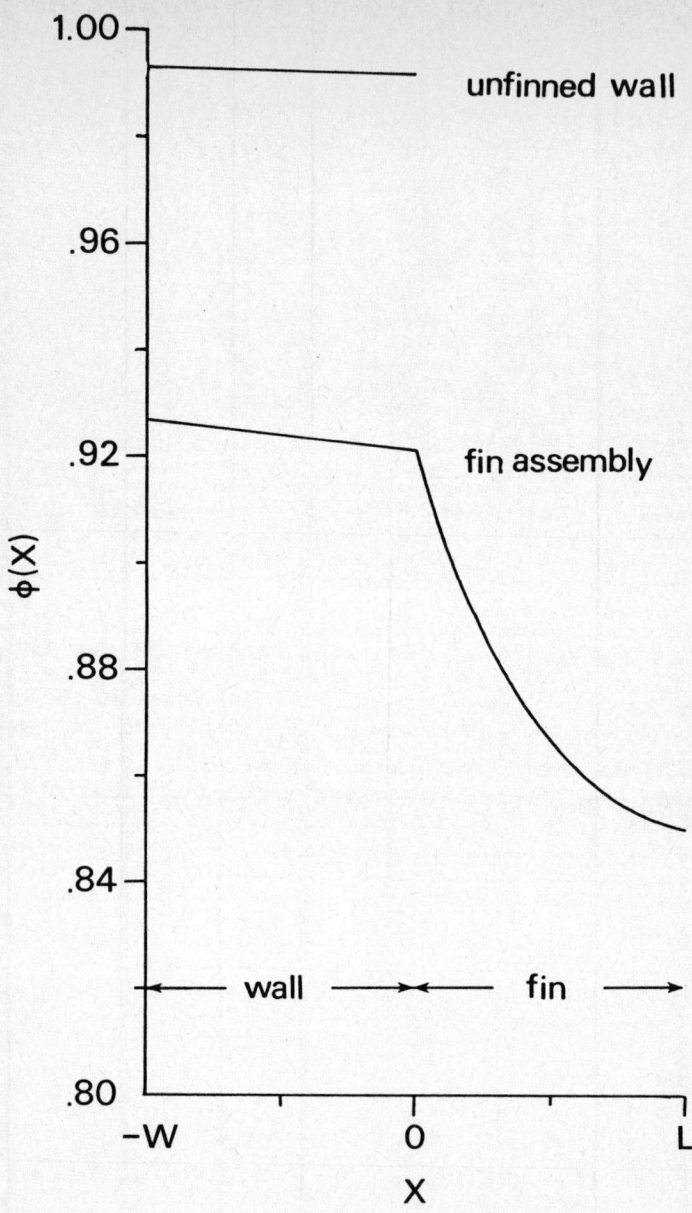

Fig.4 The one-dimensional temperature distribution for problem A

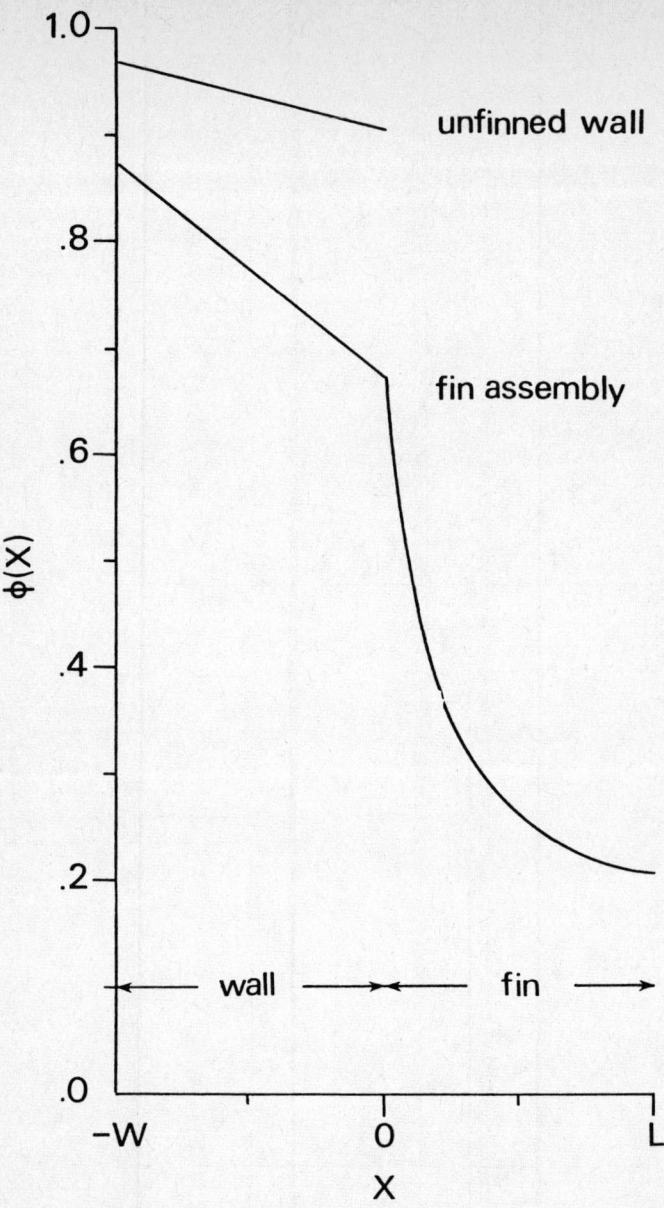

Fig.5 The one-dimensional temperature distribution for problem B

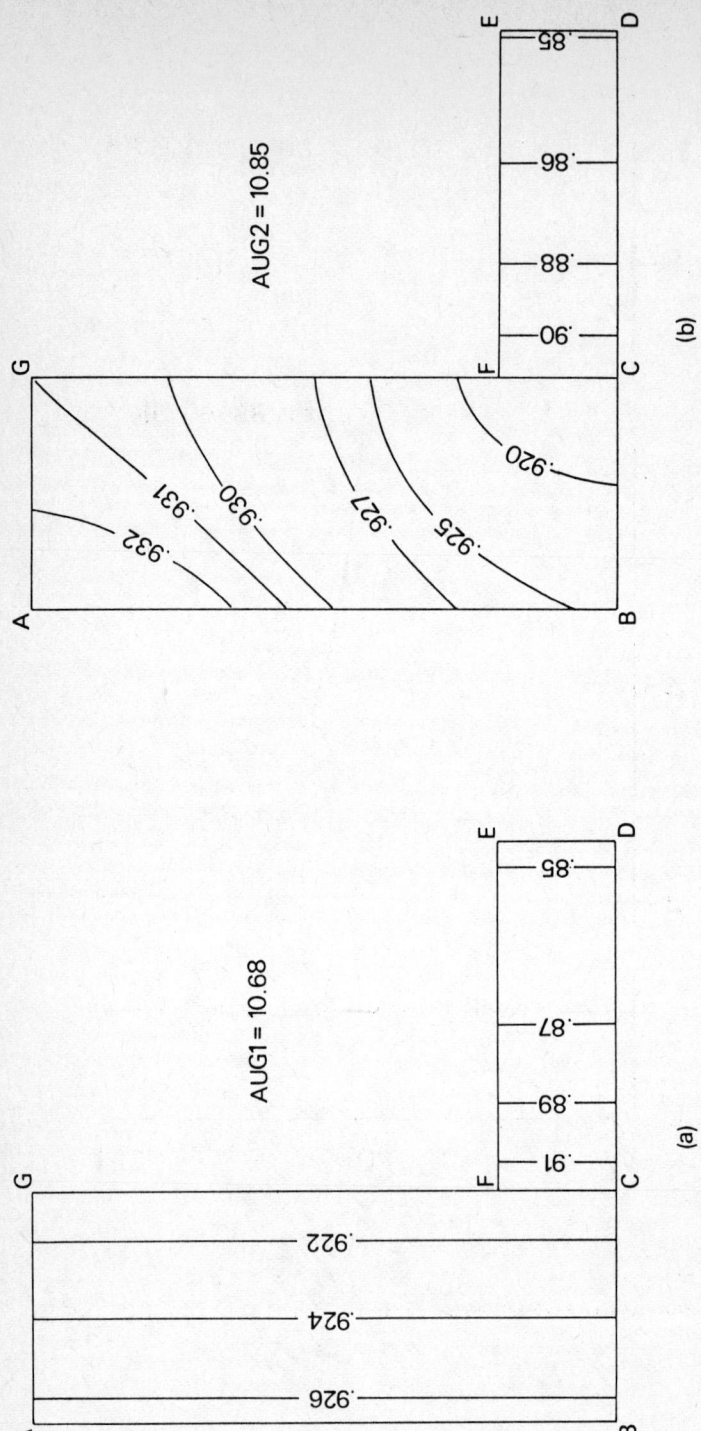

Fig.6 The one- and two-dimensional solutions for problem A

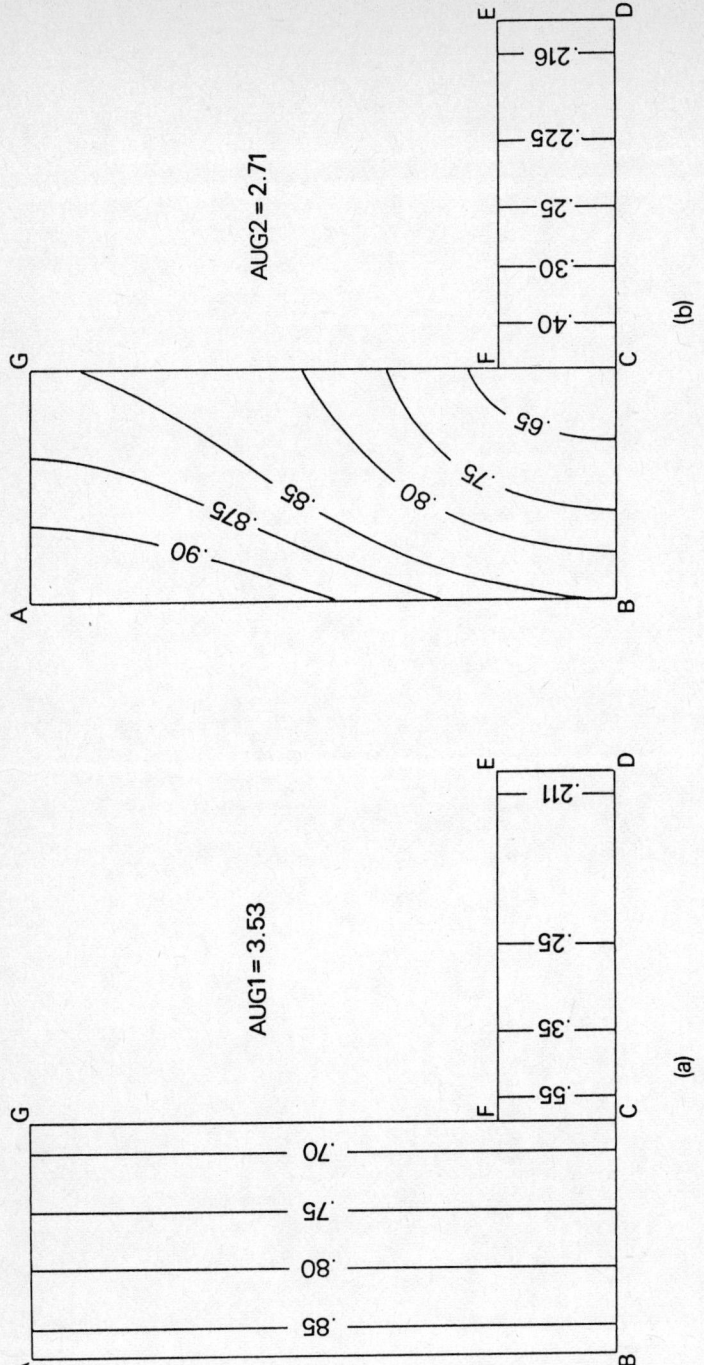

Fig.7 The one- and two-dimensional solutions for problem B

CHAPTER 5

THE APPLICABILITY OF THE PERFECT CONTACT ASSUMPTION

5.1 THE EFFECTS OF SURFACE ROUGHNESS ON THE PERFORMANCE OF EXTENDED SURFACE HEAT EXCHANGERS

ABSTRACT

The heat flow within finned surfaces is conventionally analysed on the basis that there is perfect contact between the fins and the supporting surface. However, if the fins are not an integral part of the supporting surface then the inevitable roughness of the contacting surfaces prevents contact except at discrete zones across the contact interface. In this study a simplified model of the contact interface is developed in order to investigate the applicability of the perfect contact approximation. A multi-dimensional analysis is employed in order to accurately model the heat flow and solutions are computed employing a boundary integral equation method. The results obtained indicate that it is unnecessary to consider more sophisticated representations of the contact interface.

INTRODUCTION

The analysis of the heat flow within finned heat exchangers is conventionally based upon the assumption that there is perfect contact between the fins and the supporting surface, e.g. [1,2]. This assumption facilitates a considerable reduction in the complexity of the analysis, but is strictly valid only if the fins form an integral part of the supporting surface. However, if the fins are attached to the supporting surface, e.g. by brazing, then the inevitable roughness of the contacting surfaces prevents contact except at discrete zones along the contact interface. Thus, in order to establish the applicability of present design techniques, it is essential to investigate the validity of the perfect contact approximation.

Prior to the present investigation there have only been three published articles [3,4,5] pertaining to the applicability of the perfect contact assumption. All of these articles considered the case of spiral fins wound onto a cylindrical tube and examined the deterioration in performance caused by differential expansion which completely relaxes the contact between the fins and the tube. These investigations concluded that the perfect contact analyses could over-predict the heat transfer rate by as much as 100 per cent. However, the results of these investigations are of limited value because in most applications some form of bonding is used in order to maintain contact, at least at discrete zones along the contact interface.

In contrast to the limited amount of literature pertaining to imperfect contact in fin problems, there is an extensive range of published material on the subject of thermal resistance between contacting rough surfaces, e.g. [6-12]. These studies have shown that the heat flow across the contact interface is dependent upon the distribution, shape and size of the contact zones, which in turn are governed by the mechanical, thermal and topographical properties of the contacting surfaces. The combined complexity of these features precludes an exact mathematical representation, but simplified models of the contact interface have been investigated e.g. [11,12]. Unfortunately, these models have proved to be inapplicable for the types of interfaces of practical interest. Consequently, the majority of previous investigations have been of an experimental nature, e.g.

[6-10]. Attempts to correlate the experimental data in a form convenient for design purposes have indicated discrepancies between the various investigations, e.g. [10]. This is not surprising in view of the large number of parameters and phenomena governing contact interface heat transfer.

In this study a theoretical technique is developed for predicting the effects of surface roughness on the heat transfer capabilities of finned heat exchangers. This technique does not attempt to model the complex features governing the heat transfer across the contact interface, but instead examines how prescribed contact-zone distributions affect the heat flow through the heat exchanger. The heat transfer across the gaps adjacent to the contact zones is neglected and consequently, for each particular contact zone distribution, an upper limit on the reduction in the overall heat flow is determined.

The primary objective of the present study is to investigate the validity of the perfect contact approximation in the context of fin assembly heat transfer. Attention is restricted to the case of longitudinal rectangular fins attached to a plane wall. However, the work presented here can easily be extended to the annular geometry and to include fins with tapered or curved profiles.

ANALYSIS

Assumptions and Model

Consider an assembly of equally spaced longitudinal rectangular fins attached to a plane wall, as depicted schematically in Fig. 1.

The heat flow within such systems is usually examined employing a one-dimensional analysis, e.g. [13,14]. However, several recent investigations, e.g. [1,2], have shown that these one-dimensional analyses neglect the distortion of the heat flow caused by the presence of fins. This distortion results from the fact that a greater proportion of the heat flow is channelled through the fin than through the adjacent unfinned portion of the base surface [1,2]. The consequential difference between fin assembly heat transfer rates predicted by one- and two-dimensional analyses can be as much as 80 per cent [2]. If the fins are not in perfect contact with the base surface then similar distortions, but on a much smaller scale, are caused by heat flowing through the contact zones in preference to the adjacent gaps. Thus, it is essential to employ a two-dimensional analysis in order to achieve an accurate representation of these effects.

For mathematical treatment it is necessary to develop a theoretical representation of the fin assembly. If the fins are in perfect contact with the base surface, then, by symmetry, it is only necessary to examine that section of the assembly bounded by the contour OABCDEFO, Fig. 1. However, if the fins are attached to the base-surface, then thermal symmetry is improbable as the distribution of contact zones is unlikely to be symmetric. Therefore, to facilitate a reduction in the complexity of the problem, it is assumed that the contact zones are distributed symmetrically about the axis BCD, Fig. 1. In fact, non-symmetric contact zone distributions preclude a mathematical treatment as the entire assembly would need to be considered.

The total heat flow across the contact interface is a combination of conduction across the contact zones, and conduction, convection and radiation through the gaps adjacent to the contact zones [6-12]. The convection contribution to the heat transfer across the gaps is negligible because of the microscopically small separation of the surfaces. If the fluid entrapped within the gaps has relatively high thermal conductivity, then the gap heat transfer is primarily conductive, whilst at high temperatures or in vacua, the heat transfer is mainly radiative. In this study a first approximation to the contact interface is introduced. It is assumed that there is perfect contact at the contact zones and zero heat flux across the adjacent gaps. Thus, the gap heat transfer is completely neglected. This has the effect of maximising the distortion of the heat flow and minimising the total heat transfer across the contact interface, i.e. for any given contact zone distribution, this model will give an upper bound on the respective effects of reduced wall-to-fin contact.

In comparison with the relevant dimensions of the fin assembly section OABCDEFO, the microscopic surface irregularities, which cause reduced contact, are negligible. Therefore, the contact interface OC is treated as a perfectly flat surface consisting of segments across which there is perfect contact, interspersed with segments across which there is no heat flow.

Mathematical Analysis

The determination of the fin assembly heat transfer rate requires the simultaneous solution of

$$\nabla^2 \phi_f = 0 \quad \text{within the fin,} \tag{1}$$

$$\nabla^2 \phi_w = 0 \quad \text{within the wall,} \tag{2}$$

subject to the boundary conditions

$$\text{on OA} \quad \phi_f' = -\frac{Bi_2}{\kappa} \phi_f \tag{3a}$$

$$\text{on AB} \quad \phi_f' = -\frac{Bi_2}{\kappa} \phi_f \tag{3b}$$

$$\text{on BC} \quad \phi_f' = 0 \tag{3c}$$

$$\text{on CD} \quad \phi_w' = 0 \tag{3d}$$

$$\text{on DE} \quad \phi_w' = Bi_1 (1-\phi_w) \tag{3e}$$

$$\text{on EF} \quad \phi_w' = 0 \tag{3f}$$

$$\text{on FO} \quad \phi_w' = -Bi_2 \phi_w \tag{3g}$$

$$\text{on OC} \quad \phi_f = \phi_w \text{ and } \phi_w' = -\kappa \phi_f' \quad \text{on each contact zone} \tag{3h}$$

$$\text{and} \quad \phi_f' = 0 \text{ and } \phi_w' = 0 \quad \text{on each "gap"} \tag{3i}$$

where the prime denotes the derivative in the direction of the outward normal to the associated surface.

Conditions (3c), (3d) and (3f) arise from the geometrical and thermal symmetry of the fin assembly configuration, and stipulate that there is no heat flux across the boundaries BC, CD and EF, respectively. Conditions (3h) and (3i) represent the regions of perfect contact and the adjacent gaps across which it is assumed that there is no heat transfer. The remaining conditions describe the convective heat exchange from the exposed surfaces, FOAB and DE.

The solution of the problem defined by the equations (1), (2) and (3) can be computed by various numerical techniques, e.g. the finite-difference [15], finite-element [16] and boundary integral equation (BIE) [17] methods. The BIE method is employed in this study as it can most easily handle the variable sizes of the contact zones and the complex nature of the associated contact interface boundary conditions, (3h) and (3i). In addition, the BIE method has the inherent feature that, in contrast to the finite-difference and finite-element methods, discretization for numerical purposes occurs only on the domain boundary and therefore generates a considerably smaller system of equations than an equivalent finite-difference or finite-element approximation. Thus, in comparison with the finite-difference and finite-element methods, the BIE formulation facilitates substantial reductions in the computational storage and time requirements.

For the particular fin assembly geometry considered here, solutions can also be computed by a very sophisticated numerical implementation of the separation of variables method [18]. For the solution of problems involving perfect wall-to-fin contact this separation of variables method is computationally more efficient than the BIE method, i.e. the separation of variables method gives more accurate solutions but requires less computational storage and time than the BIE method [18]. However, the extension of this separation of variables method to incorporate the idealised imperfect contact interface, proposed in the preceding section, involves substantial increases in both the programming effort and the computational requirements. This is a consequence of an increased complexity in the analysis. In contrast, the extension of the BIE method, to problems involving non-perfect wall-to-fin contact, is achieved with appreciably little additional programming effort and a minimal increase in the computational requirements. Nevertheless, the separation of variables method has been implemented for the 1 and 2 contact zone cases in order to verify the accuracy of the BIE method; the solutions obtained employing the separation of variables method are in exact agreement with those predicted by the BIE method.

FIN ASSEMBLY HEAT TRANSFER RATE

The heat flow rate through the fin assembly is most conveniently characterised in the form of an augmentation factor, Aug, defined as the ratio of the heat transfer rate of the finned assembly to that of the unfinned wall operating under the same conditions. This augmentation factor can be evaluated at either of the exposed surfaces DE and FOAB, and is given by

$$\text{Aug} = \left(\frac{1}{Bi_1} + W + \frac{1}{Bi_2}\right) \int_{DE} \phi_w'(q)\,dq \qquad (4)$$

$$= -\left(\frac{1}{Bi_1} + W + \frac{1}{Bi_2}\right)\left\{\int_{FO} \phi_w'(q)\,dq + \kappa \int_{OAB} \phi_f'(q)\,dq\right\}. \qquad (5)$$

In the context of the BIE solutions these integrations can be performed numerically in a manner consistent with the overall discretization. However, as the BIE solution is only an approximation, it need not give exactly the same values for the integrations (4) and (5), although for the solutions to be satisfactory these should agree within an acceptable tolerance.

RESULTS AND DISCUSSION

From the equations (1), (2) and (3) it may be deduced that the heat flow through the fin assembly can be parameterised by the Biot numbers Bi_1 and Bi_2, the ratio of the thermal conductivities κ, the aspect ratios L, T and W, and the size and distribution of the contact zones. In order to minimise the number of parameters necessary to describe the fin assembly, a systematic procedure is adopted for examining the effects of variations in the size and distribution of the contact zones. The contact interface is parameterised by a single parameter, namely ε, which indicates the ratio of the actual contact area to the apparent contact area. Then, for each prescribed value of ε, solutions are computed for four different contact zone distributions. These correspond to the cases of 1, 2, 4 and 8 contact zones as shown in Fig. 2. Thus, variations in both contact area and contact zone distributions are accounted for by the single parameter ε.

Results have been obtained for a wide range of the system parameters Bi_1, Bi_2, κ, L, T and W. For each particular problem, solutions were computed for the cases ε = 0.0, 0.1, 0.2, 0.5 and 1.0, i.e. with the amount of contact varying from zero to complete contact. The results for 3 particular problems are presented in Tables, 1, 2 and 3. These tables show the values of the augmentation factor corresponding to the various values of ε and the different contact zone distributions. These results correspond to the problems

1. Bi_1 = 1.00, Bi_2 = 0.01, κ = 1.0, L = 10.0, T = 0.5 and W = 5.0

2. Bi_1 = 1.00, Bi_2 = 0.01, κ = 10.0, L = 10.0, T = 0.5 and W = 5.0

3. Bi_1 = 0.10, Bi_2 = 0.001, κ = 2.0, L = 10.0, T = 0.5 and W = 5.0

Problem 1 represents the performance of a stainless steel finned heat exchanger with forced convection air flow on the plain side and free convection air flow on the fin side. Problem 2 corresponds to a heat exchanger comprised of a stainless steel wall with aluminium fins, while problem 3 relates to the case of copper fins attached to an aluminium wall.

The results presented in Tables 1, 2 and 3 illustrate various features of reduced wall-to-fin contact and are characteristic of those observed for other values of the system parameters. The dominant feature of these results is that, irrespective of the distribution of contact zones, provided actual contact occurs over at least 10 per cent of the apparent contact area, then the reduction in augmentation is virtually negligible. However, if the contact area is reduced below 10 per cent then the heat exchanger performance begins to deteriorate. It has not been possible to establish the exact manner of this deterioration because the solutions within the range $0 < \varepsilon < 0.1$ become susceptible to large errors.

Closer inspection of the results displayed in Tables 1, 2 and 3 reveals that for a given amount of actual contact area, i.e. for a prescribed value of ε, increasing the number of contact zones facilitates an improvement in the fin assembly heat transfer rate even though the size of the individual contact zones is reduced. A reduction in the size of the contact zones should cause a constriction to the heat flow. However, these smaller contact zones are distributed over a wider portion of the contact interface and therefore counteract the constriction by enabling the heat flow to disperse over a larger area. To emphasise this point solutions have been computed by reducing the fin-base thickness from T to εT and then assuming perfect contact between wall and fins. These solutions are displayed in Table 4. Comparison with the results in Tables 1, 2 and 3 shows that the reductions in the fin assembly heat transfer caused by reductions in fin-base thickness are substantially greater than the reductions caused by non-perfect contact, i.e. the constriction to heat flow caused by the fin-base far outweighs that caused by the non-contacting portions of the fin-base with the wall.

In order to ensure that the phenomena observed are not a consequence of the particular contact zone distributions employed in this study, solutions have also been computed for various other contact zone distributions. These results substantiated the effects discussed above.

The effects of the reduced contact, on the fin assembly performance, predicted by the model developed in this study, are in complete contrast to those observed in situations where the contacting surfaces have the same dimensions, e.g. [6-12]. This apparent discrepancy is easily explained. In the fin assembly situation the overall heat flow is governed primarily by the constriction caused by the presence of the fins; as explained above, the constriction caused by imperfect wall-to-fin contact is comparatively negligible. However, if the contacting surfaces have the same dimensions, then the only constriction to the heat flow results from the imperfect contacting of the surfaces and consequently the overall heat flow is controlled by the contact zones.

CONCLUSIONS

A practical method for investigating the effects of reduced wall-to-fin contact, on the peformance of finned heat exchangers, has been developed. This method introduces a relatively simple representation of surface roughness, in which the heat transfer across the gaps adjacent to the contact zones is neglected. Since, in practice, there will be some heat transfer across these gaps, it is apparent that this representation of the contact interface leads to an underestimation of the heat exchanger performance. The results obtained indicate that if the actual contact area is greater than 10 per cent of the apparent contact area, then the difference between the partial and perfect contact solutions is virtually negligible, i.e. the perfect contact approximation is applicable provided actual contact occurs over at least 10 per cent of the wall-to-fin interface. It may therefore be concluded that it is only necessary to consider the gap heat transfer if the contact is less than 10 per cent.

NOMENCLATURE

Aug	general augmentation factor
Bi_1	$= h_1 P/k_w$, Biot number
Bi_2	$= h_2 P/k_w$, Biot number
h	heat transfer coefficient, $W/m^2 K$
k	thermal conductivity, W/mK
l	fin length, m
L	$= l/P$, aspect ratio
P	half fin-pitch, m
t	half fin thickness, m
T	$= t/P$, aspect ratio
w	wall thickness, m
W	$= w/P$, aspect ratio
ε	= actual contact area/apparent contact area
κ	$= k_f/k_w$
θ_1, θ_2	fluid temperatures, K
θ	temperature distribution, K
ϕ	$= (\theta-\theta_2)/(\theta_1-\theta_2)$, dimensionless temperature distribution

Subscripts

1	plain side
2	fin side
f	fin
w	wall

REFERENCES

1. Sparrow, E.M. and Lee, L., "Effects of fin-base temperature depression in a multifin array", Journal of Heat Transfer, Vol. 97, 1975, pp. 463-465.

2. Suryanarayana, N.V., "Two-dimensional effects on heat transfer rates from an array of straight fins", Journal of Heat Transfer, Vol. 99, 1977, pp. 129-132.

3. Gardner, K.A. and Carnovos, T.C., "Thermal contact resistance in finned tubing", Journal of Heat Transfer, Vol. 82, 1960, pp. 279-293.

4. Young, E.H. and Briggs, D.E., "Bond resistance of bi-metallic finned tubes," Chemical Engineering Progress, Vol. 61, 1965, pp. 71-84.

5. Kulharni, M.V. and Young, E.H., "Bi-metallic finned tubes", Chemical Engineering Progress, Vol. 62, 1966, pp. 69-73.

6. Williams, A., "Heat flow across stacks of steel laminations", Journal of Mechanical Engineering Science, Vol.13, 1971, pp. 217-223.

7. Thomas, T.R. and Probert, S.D., "Correlations for thermal contact conductance in vacua", Journal of Heat Transfer, Vol. 94, 1972, pp.276-28

8. Madhusudana, C.V., "The effect of interface fluid on thermal contact conductance", International Journal of Heat and Mass Transfer, Vol. 18, 1975, pp. 989-991.

9. O'Callaghan, P.W., Jones, A.M. and Probert, S.D., "Effect of thermal contact resistance on the performance of transfomer lamination stacks", Applied Energy, Vol.3, 1977, pp.13-22.

10. Edmonds, M.J., Jones, A.M., O'Callaghan, P.W. and Probert, S.D., "The prediction and measurement of thermal contact resistance", Wear, Vol. 50, 1978, pp. 299-319.

11. Barber, J.R., "The effect of thermal distortion on constriction resistance", International Journal of Heat and Mass Transfer, Vol. 14, 1971, pp. 751-766.

12. Mikic, B.B., "Thermal contact conductance: Theoretical considerations, International Journal of Heat and Mass Transfer", Vol. 17, 1974, pp. 205-214.

13. Gardner, K.A., "Efficiency of extended surface", Transactions of the ASME, Vol. 67, 1945, pp. 621-631.

14. Mikk, I., "Convective fin of minimum mass", International Journal of Heat and Mass Transfer, Vol. 23, 1980, pp. 707-711.

15. Smith, G.D., Numerical solution of partial differential equations, Oxford University Press, 1974.

16. Zienkiewicz, O.C., The finite element method in engineering, McGraw-Hill, London, 1971.

17. Jaswon, M.A. and Symm, G.T., Integral equation methods in potential theory and electrostatics, Academic Press, London, 1977.

18. Heggs, P.J., Ingham, D.B. and Manzoor, M., "The analysis of fin assembly heat transfer by a series truncation method", accepted for publication in Journal of Heat Transfer, 1981.

Table 1 : The effects of reduced contact for Problem 1

Number of
Contact Zones   Augmentation Factor

	$\varepsilon=0.0$	$\varepsilon=0.1$	$\varepsilon=0.2$	$\varepsilon=0.5$	$\varepsilon=1.0$
1	0.51	4.62	4.77	4.99	5.11
2	0.51	4.74	4.88	5.05	5.11
4	0.51	4.98	5.02	5.09	5.11
8	0.51	5.06	5.08	5.10	5.11

Table 2 : The effects of reduced contact for Problem 2

Number of
Contact Zones   Augmentation Factor

	$\varepsilon=0.0$	$\varepsilon=0.1$	$\varepsilon=0.2$	$\varepsilon=0.5$	$\varepsilon=1.0$
1	0.51	9.14	9.19	9.25	9.29
2	0.51	9.17	9.21	9.26	9.29
4	0.51	9.25	9.26	9.28	9.29
8	0.51	9.27	9.28	9.28	9.29

Table 3 : The effects of reduced contact for Problem 3

Number of
Contact Zones   Augmentation Factor

	$\varepsilon=0.0$	$\varepsilon=0.1$	$\varepsilon=0.2$	$\varepsilon=0.5$	$\varepsilon=1.0$
1	0.50	6.17	6.30	6.52	6.66
2	0.50	6.39	6.49	6.61	6.66
4	0.50	6.55	6.59	6.64	6.66
8	0.50	6.61	6.63	6.66	6.66

Table 4 : The results for Problems 1, 2 and 3 assuming perfect wall-to-fin contact with reduced fin thickness

Problem   Augmentation Factor   ($T = \varepsilon T$)

	$\varepsilon=0.0$	$\varepsilon=0.1$	$\varepsilon=0.2$	$\varepsilon=0.5$	$\varepsilon=1.0$
1	1.00	2.67	3.31	4.35	5.11
2	1.00	7.46	8.37	9.04	9.29
3	1.00	4.93	5.63	6.33	6.66

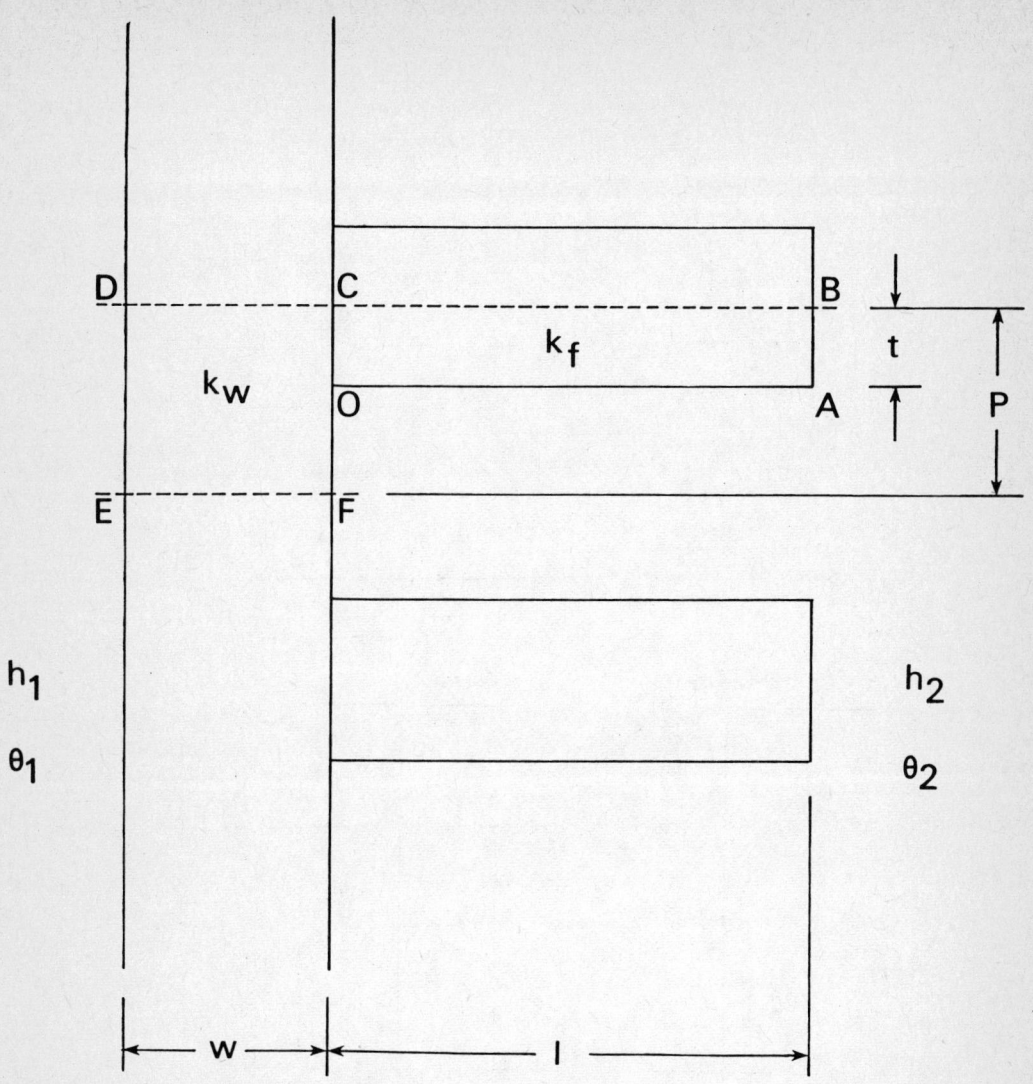

Fig.1 Schematic representation of the fin assembly

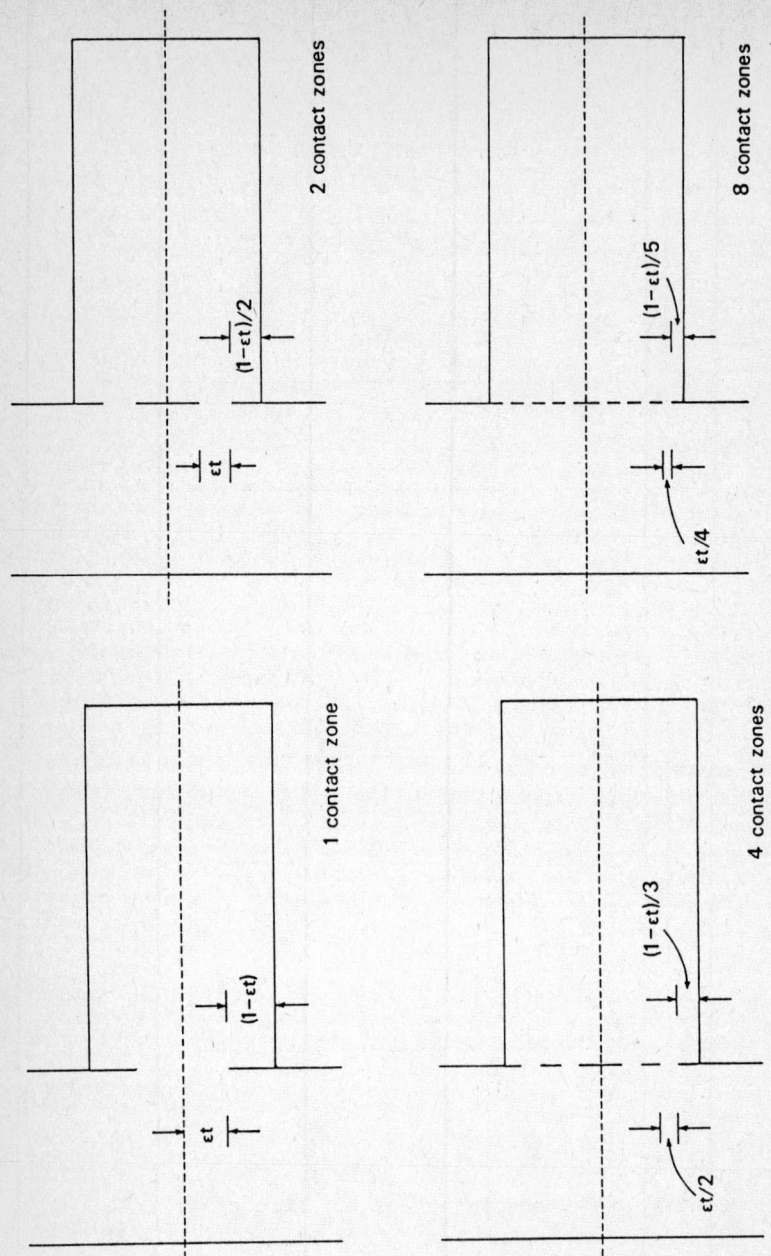

Fig.2 Contact zone distributions

5.2 THE EFFECTS OF INTERFACIAL BONDING ON THE PERFORMANCE OF EXTENDED SURFACE HEAT EXCHANGERS

ABSTRACT

The heat flow within finned surfaces is conventionally analysed on the basis that there is perfect contact between the fins and the supporting surface. However, in many applications the fins are attached to the supporting surface by some form of bonding, e.g. brazing, soldering or welding, but the presence of the bonding material is not accounted for by the perfect contact analyses. In this study a simple theoretical representation of interfacial bonding is developed in order to investigate the applicability of the perfect contact approximation. A multi-dimensional analysis is employed in order to accurately model the heat flow and solutions are computed employing a boundary integral equation method. The results obtained indicate, in particular, that it is unnecessary to consider more sophisticated representations of interfacial bonding.

INTRODUCTION

The analysis of the heat flow within finned heat exchangers is conventionally based upon the assumption that there is perfect contact between the fins and the supporting surface, e.g. [1,2]. This assumption facilitates a considerable reduction in the complexity of the analysis, but is strictly valid only if the fins form an integral part of the supporting surface. However, in most applications the fins are bonded to the supporting surface, e.g. by brazing, soldering or welding, and this introduces some additional material, namely the bonding material, between the fins and the supporting surface. The presence of this bonding material is not accounted for in the perfect contact analyses. Thus, in order to establish the applicability of present design techniques, it is essential to investigate the validity of the perfect contact approximation.

Prior to the present investigation there have only been three published articles [3,4,5] pertaining to the applicability of the perfect contact assumption. All of these articles considered the case of spiral fins wound onto a cylindrical tube and examined the deterioration in performance caused by differential expansion which completely relaxes the contact between the fins and the tube. These investigations concluded that the perfect contact analyses can overpredict the heat transfer rate by as much as 100 per cent. However, the results of these investigations are of limited value because in most applications the bonding material maintains contact, at least at discrete zones along the contact interface.

In contrast to the limited amount of literature pertaining to imperfect in fin problems, there is an extensive range of published material on the subject of thermal resistance between contacting surfaces of the same dimensions, e.g. [6-12]. These studies have shown that the heat flow across the contact interface is dependent upon the mechanical, thermal and topographical properties of the components which constitute the contact interface. The combined complexity of these features precludes an exact mathematical representation of the contact interface. Simple models have been developed but these are only applicable for directly contacting surfaces, i.e. in the absence of any interfacial inserts e.g. [6,7,8]. The majority of the previous work concerning heat transfer across joints which include an insert has been experimental, e.g. [9-12]. Attempts to correlate this data in a form convenient for design purposes have shown discrepancies between the results of the various investigations, e.g. [10]. This is not surprising in view of the large number of parameters and phenomena governing the interfacial heat transfer.

In this study a theoretical technique is developed for predicting the effects of bonding on the heat transfer capabilities of finned heat exchangers. This technique does not attempt to model the complex features governing contact, but instead examines how variations in the conductivity, thickness and penetration of the bonding material affect the heat flow through the heat exchanger. In situations where the bonding material does not penetrate across the entire contact interface, the heat transfer across the cavity included between the bonding material is neglected and consequently an upper limit on the reduction in the overall heat flow is determined.

The primary objective of the present study is to investigate the applicability of the perfect contact approximation in the context of fin assembly heat transfer. Attention is restricted to the case of longitudinal rectangular fins attached to a plane wall. However, the work presented here can easily be extended to the annular geometry and to include fins with curved or tapered profile.

ANALYSIS

Assumptions

Consider a heat exchanger comprised of equally spaced longitudinal rectangular fins attached to a plane wall. In order to enable a mathematical treatment it is necessary to develop a theoretical representation of this device. If there is perfect contact between the fins and the base-surface, then the inherent thermal symmetry permits the fin assembly to be represented by the region ABCDEFGA, Fig. 1, [2]. However, if the fins are bonded to the base-surface then thermal symmetry occurs only if the bonding material is distributed symmetrically about the axis BCD, Fig. 1. Consequently, non-symmetrical distributions of the bonding material are not susceptible to a mathematical treatment because they necessitate the examination of the entire fin assembly. Therefore, in the following analysis attention is restricted to symmetrical distributions of the bonding material. From a practical viewpoint this restriction is not unrealistic because the bonding material is usually applied in some symmetrical configuration, e.g. as depicted schematically in Fig. 2. The type of contact configuration actually employed in any particular situation will be governed primarily by the conditions under which the finned surface is required to operate, e.g. high pressure drop conditions demand a relatively strong contact such as that afforded by configuration (b), Fig. 2, whilst in a free convection environment, configuration (a), Fig. 2, would suffice.

The attachment of the fins to the base-surface can also be achieved without actually inserting the bonding material between the fin-base and the base-surface, e.g. as shown in Fig. 3. In such situations actual contact of the fin-base to the base-surface only occurs at discrete zones across the contact interface FC, Fig. 3, [8]. This is due to the inevitable roughness of the contacting surfaces [8]. The analysis of such contacts has been examined in detail by Heggs et al [8] and therefore, in this study, attention will be restricted to contacts in which the presence of the bonding material physically separates the fin-base from the base-surface.

For contacts of the type shown in Fig. 2, the total heat flow across the contact interface is a combination of conduction across the bridges formed by the bonding material, and conduction, convection and radiation through the cavity included between the contact bridges, [10]. If the cavity is in vacuum then the cavity heat transfer will be entirely radiative, [10]. However, if the cavity contains a fluid then the heat transfer will be primarily conductive, although, at high temperatures or for low conductivity fluids, radiation may be significant,[10]; the convection contribution to the cavity heat transfer is negligible because of the enclosed nature of the cavity, [10]. In this study a first approximation to the contact interface is introduced. It is assumed that there is no heat flux across the cavity. This assumption is introduced primarily to facilitate a reduction in the complexity of the analysis. However, the associated results are very useful for design purposes; neglecting the cavity heat transfer has the effect of minimising the total heat transfer across the contact interface and therefore gives rise to an upper-bound on the reduction in the overall heat flow through the fin assembly caused by the presence of the bonding material.

Mathematical Analysis

The primary objective of the present study is to investigate the validity of the perfect contact approximation in situations where fin-to-base-surface contact is achieved by the introduction of some additional material. In order to facilitate a concise mathematical treatment, whilst still retaining the essential features of the actual physical situation, attention is restricted to contacts of the form indicated by configuration (a) in Fig. 2. More sophisticated contact configurations necessitate more independent parameters in order to describe the contact interface and, in addition, involve a considerably more complicated mathematical analysis.

The analysis of the heat flow within fin system is conventionally based upon the assumption that the heat flow is one-dimensional, [1]. However, recent investigations have shown that these one-dimensional analyses neglect the distortion of the heat flow within the assembly caused by the presence of fins, e.g. [13,14]. These distortions result from the fact that a greater proportion of the heat flow is channelled through the fin than through the adjacent unfinned portion of the base-surface, e.g. [13,14]. Suryanarayana [14] has reported that the difference between fin assembly heat transfer rates predicted by one- and two-dimensional analyses can be as much as 80 per cent. If the fins are attached to the base-surface by the introduction of some additional material, as indicated in Fig. 2, then similar distortions are caused by the heat flowing through the bonding material in preference to the adjacent cavity. Thus, it is essential to employ a two-dimensional analysis in order to achieve an accurate representation of these effects.

For the fin assembly configurations shown in Fig. 4, the determination of the fin assembly heat transfer rate requires the simultaneous solution of

$$\nabla^2 \phi_w = 0 \quad \text{within the wall,} \tag{1}$$

$$\nabla^2 \phi_b = 0 \quad \text{within the bonding material,} \tag{2}$$

and

$$\nabla^2 \phi_f = 0 \quad \text{within the fin,} \tag{3}$$

subject to the boundary conditions

on AB	$\phi_w' = Bi_1(1 - \phi_w)$	(4a)
on BC	$\phi_w' = 0$	(4b)
on CD	$\phi_w' = 0$	(4c)
on DE	$\phi_b' = 0$	(4d)
on EF	$\phi_f' = 0$	(4e)
on FG	$\phi_f' = 0$	(4f)
on GH	$\phi_f' = -\dfrac{Bi_2}{\kappa_f}\phi_f$	(4g)
on HI	$\phi_f' = -\dfrac{Bi_2}{\kappa_f}\phi_f$	(4h)
on IJ	$\phi_b' = -\dfrac{Bi_2}{\kappa_b}\phi_b$	(4i)
on JK	$\phi_w' = -Bi_2\,\phi_w$	(4j)
on KA	$\phi_w' = 0$	(4k)
on IE	$\phi_b = \phi_f$	(4l)
and	$\phi_b' = -\dfrac{\kappa_f}{\kappa_b}\phi_f'$	(4m)
on DJ	$\phi_w = \phi_b$	(4n)
and	$\phi_w' = -\kappa_b\,\phi_b'$	(4o)

where the prime (') denotes the derivative in the direction of the outward normal to the associated surface.

Conditions (4b), (4f) and (4k) arise from the thermal and geometrical symmetry of the fin assembly configuration and stipulate that there is no heat flux across the boundaries BC, FG and KA, respectively. Conditions (4c), (4d) and (4e) result from neglecting the cavity heat transfer, whilst conditions (4l), (4m), (4n) and (4o) indicate perfect contact of the bonding material to both the fins and the base-surface. The remaining boundary conditions describe the convective heat exchange from the exposed surfaces, AB and GHIJK.

The problem described by equations (1), (2), (3) and (4) is susceptible to treatment by various numerical techniques, e.g. the finite difference [15], finite-element [16] and boundary integral equation (BIE) [17] methods. The BIE method is employed in this study as it can most easily handle the different sizes of the various regions and the complex nature of the associated boundary conditions. In addition, the BIE method has the inherent feature that, in contrast to the finite-difference and finite-element methods, discretization for numerical purposes occurs only on the domain boundary and therefore generates a considerably smaller algebraic representation than an equivalent finite-difference or finite-element approximation. Thus, in comparison with the finite-difference and finite-element methods, the BIE formulation facilitates substantial reductions in the computational storage and time requirements.

FIN ASSEMBLY HEAT TRANSFER RATE

The heat flow through the fin assembly is most conveniently characterised in the form of an augmentation factor, Aug, defined as the ratio of the heat transfer rate of the finned assembly to that of the unfinned wall operating under the same conditions. This augmentation factor can be evaluated at either of the exposed surfaces AB or GHIJK, Fig. 4, and is given by

$$\text{Aug} = \left(\frac{1}{Bi_1} + W + \frac{1}{Bi_2}\right) \int_{AB} \phi'_w(q)\,dq \tag{5}$$

$$= -\left(\frac{1}{Bi_1} + W + \frac{1}{Bi_2}\right) \left\{ \kappa_f \int_{GHI} \phi'_f(q)\,dq + \kappa_b \int_{IJ} \phi'_b(q)\,dq \right.$$

$$\left. + \int_{JK} \phi'_w(q)\,dq \right\} \tag{6}$$

In the context of the BIE solutions these integrations can be performed numerically in a manner consistent with the overall discretization. However, as the BIE solution is only an approximation, it need not give exactly the same values for the expressions (5) and (6), although, for the solution to be satisfactory, these should agree to within an acceptable tolerance.

RESULTS AND DISCUSSION

The heat flow through the fin assembly is parameterised by the Biot numbers Bi_1 $(= h_1 P/k_w)$ and Bi_2 $(= h_2 P/k_w)$, the dimensionless thermal conductivities κ_b $(= k_b/k_w)$ and κ_f $(= k_f/k_b)$, the aspect ratios L $(= l/P)$, $T (= t/P)$ and W $(= w/P)$ and the parameters δ and ε which describe the dimensions of the bond region as percentages of the fin length and fin-base thickness, Fig. 4. The parameters Bi_1, Bi_2, κ_f, L, T and W define the basic fin assembly, and the parameters κ_b, δ and ε describe the contact interface. Solutions have been computed for a wide range of the system parameters. In order to investigate the effects of bonding on the overall heat flow through the fin assembly, for each particular problem the parameters Bi_1, Bi_2, κ_f, L, T and W were assigned values and then solutions were computed for a comprehensive range of the parameters κ_b, δ and ε. The results for two particular problems are presented in Figures 5 and 6. These figures show the behaviour of the augmentation factor, Aug, with variations in the contact interface parameters κ_b, δ and ε. The results presented in Figs. 5 and 6 correspond respectively to the problems:

A : $Bi_1 = 1.0$, $Bi_2 = 0.010$, $\kappa_f = 10.0$, $L = 10.0$, $T = 0.5$ and $W = 5.0$,

B : $Bi_1 = 0.1$, $Bi_2 = 0.001$, $\kappa_f = 2.0$, $L = 10.0$, $T = 0.5$ and $W = 5.0$.

Problem A is representative of a heat exchanger comprised of a stainless-steel wall with aluminium fins, and Problem B relates to a heat exchanger consisting of an aluminium wall with copper fins. For both problems the effects of variations in the thermal conductivity of the bonding material have been investigated by computing solutions first for the case of a low conductivity lead-tin alloy and then for a high conductivity copper-silver-zinc alloy; Figs. 5a and 6a relate to the low conductivity bonding material and Figs. 5b and 6b to the high conductivity bonding material. The effects of variations in the thickness and penetration of the contact region DEIJ, Fig. 4, have been investigated by determining solutions for δ ranging from 0 to 10 per cent and for ε ranging from 0 to 100 percent, i.e. for the thickness of the contact region varying from 0 to 10 per cent of the fin length and for the penetration varying from 0 to 100 per cent of the contact interface. However, the solutions for values of ε less than 10 per cent were found to involve relatively large errors and therefore have not been included in Figs. 5 and 6.

With reference to Figs. 5 and 6, it should be noted that the case $\delta = 0$ corresponds to the situation in which there is no bonding material present at the contact interface and therefore relates to the two-dimensional perfect-contact solution, Aug^*.

The results presented in Figs. 5 and 6 illustrate various effects of the bond resistance and are characteristic of those observed for other values of the system parameters. It is clearly evident from these results that reductions in the penetration cause a marked deterioration in the performance of the fin assembly. This deterioration in performance is due to the fact that for smaller penetrations the area through which the heat flow is channelled is smaller and consequently the constriction to the heat flow is greater. In order to substantiate this phenomena, solutions have been computed to problem A by reducing the fin-base thickness from T to εT and then assuming perfect wall-to-fin contact:

ε	Aug^* (for $T = \varepsilon T$)
100%	6.66
50%	6.33
20%	5.63
10%	4.93

When the fin-base thickness is reduced from the prescribed value to 10 per cent of this value, the reduction in the overall fin-side heat transferring surface area is less than 5 per cent but the respective reduction in the augmentation is over 25 per cent. Thus, the constriction to the heat flow caused by the fin-base, i.e. the area through which the heat flow is channelled, controls the overall heat flow through the fin assembly.

The results presented in Figs. 5 and 6 indicated that the heat exchanger performance deteriorates if the thickness of the bonding material is increased, but improves if the thermal conductivity of the contact is increased. These phenomena are both related to the conductive resistance of the contact; increases in the thickness of the contact cause an increase in the conductive resistance, whilst increases in the thermal conductivity of the contact facilitate a reduction in the conductive resistance.

In the range $\delta < 5$ per cent and $\varepsilon < 50$ per cent, the modified fin assembly model developed in this study indicates that, for realistic bonding materials, the maximum reduction in the fin assembly heat transfer rate, from that predicted by the conventional perfect-contact analyses, is less than 5 per cent, e.g. see Figs. 5 and 6. However, as this model neglects the heat transfer across cavities in the contact interface, it is apparent that the actual heat exchanger performance will be greater than that predicted here. Thus, from a design viewpoint, these results suggest that, if possible, the bonding material should be applied such that its thickness is less than 5 per cent of the fin length and that it achieves at least 50 percent penetration across the contact interface because under such circumstances the bond resistance is virtually negligible.

It has not been possible to obtain any data pertaining to the actual dimensions of the contact region in practical heat exchangers, however, intuitively it appears that the requirements that δ be less than 5 per cent and ε be greater than 50 per cent, can be achieved without difficulty.

CONCLUSIONS

A practical method has been developed for investigating the effects of bond resistance on the performance of finned heat exchangers. This method introduces a relatively simple representation of the contact interface in which the heat transfer across the cavities in the contact interface is completely neglected. Since, in practice, there will be some heat transfer across such cavities, it is apparent that this representation leads to an under-estimation of the heat exchanger performance. The results obtained indicate that for realistic contacts the difference between the solutions predicted by the model developed in this study and those predicted by the two-dimensional perfect-contact analysis is virtually negligible.

NOMENCLATURE

Aug   general augmentation factor

Aug^* perfect-contact augmentation factor

Bi_1 $= h_1 P/k_w$, Biot number

Bi_2 $= h_2 P/k_w$, Biot number

h   heat tranfer coefficient, $W/m^2 K$

k   thermal conductivity, W/mK

l   fin length, m

L $= l/P$, aspect ratio

P   half fin-pitch, m

t   half fin thickness, m

T $= t/P$, aspect ratio

w   wall thickness, m

W $= w/P$, aspect ratio

δ thickness of the bond as a percentage of the fin length

ε penetration of the bond as a percentage of half the fin-base thickness.

κ_b $= k_b/k_w$

κ_f $= k_f/k_w$

θ_1, θ_2 fluid temperatures, K

θ temperature distribution, K

ϕ $= (\theta - \theta_2)/(\theta_1 - \theta_2)$, dimensionless temperature distribution

Subscripts

1   plain side

2   fin side

b   bond

f   fin

w   wall

REFERENCES

1. I. Mikk, "Convective fin of minimum mass", International Journal of Heat and Mass Transfer, Vol. 23, pp. 707-711, 1980.

2. P.J. Heggs and P.R. Stones, "The effects of dimensions on the heat flow rate through extended surfaces", Journal of Heat Transfer, Vol. 102, pp. 180-182, 1980.

3. Gardner, K.A. and Carnovos, T.C., "Thermal contact resistance in finned tubing", Journal of Heat Transfer, Vol. 82, 1960, pp. 279-293.

4. Young, E.H. and Briggs, D.E., "Bond resistance of bi-metallic finned tubes", Chemical Engineering Progress, Vol. 61, 1965, pp. 71-84.

5. Kulkarni, M.V. and Young, E.H., "Bi-metallic finned tubes", Chemical Engineering Progress, Vol. 62, 1966, pp. 69-73.

6. B.B. Mikic, "Thermal contact conductance: theoretical considerations", International Journal of Heat and Mass Transfer, Vol. 17, pp. 205-214, 1974.

7. C.V. Madhusudana, "The effect of interface fluid on thermal contact conductances", International Journal of Heat and Mass Transfer, Vol. 18, pp. 989-991, 1975.

8. P.J. Heggs, D.B. Ingham and M. Manzoor, "The effects of thermal contact resistance on the performance of finned heat exchangers", To be presented at the Second National Symposium on Numerical Methods in Heat Transfer, University of Maryland, USA, September 1981.

9. R.P. Forslund, "An experimental technique for determining the diffusion effect of braze material on the thermal conductance of thin copper fins", Journal of Heat Transfer, Vol. 25, pp. 273-275, 1972.

10. P.W. O'Callaghan, A.M. Jones and S.D. Probert, "Research Note: The thermal behaviour of gauzes as interfacial inserts between solids", Journal of Mechanical Engineering Science, Vol. 17, pp. 233-236, 1975.

11. F.R. Al-Astrabadi, P.W. O'Callaghan, A.M. Jones and S.D. Probert, "Thermal resistance resulting from commonly used inserts between stainless steel static bearing surfaces", Wear, Vol. 40, pp. 339-350, 1976.

12. F.R. Al-Astrabadi, S.D. Probert, P.N. O'Callaghan and A.M. Jones, "Reduction of energy dissipation at thermally distorted pressed contacts", Applied Energy, Vol. 5, pp. 23-51, 1979.

13. E.M. Sparrow and L. Lee, "Effects of fin-base temperature depression in a multifin array", Journal of Heat Transfer, Vol. 197, pp. 463-465, 1975.

14. N.V. Suryanarayana, "Two-dimensional effects on heat transfer rates from an array of straight fins", Journal of Heat Transfer, Vol. 99, pp. 129-132, 1977.

15. G.D. Smith, Numerical Solution of Partial Differential Equations, Oxford University Press, 1974.

16. O.C. Zienkiewicz, The Finite Element Method in Engineering, McGraw-Hill, London 1971.

17. M.A. Jaswon and G.T. Symm, Integral Equation Methods in Potential Theory and Electrostatics, Academic Press, London, 1977.

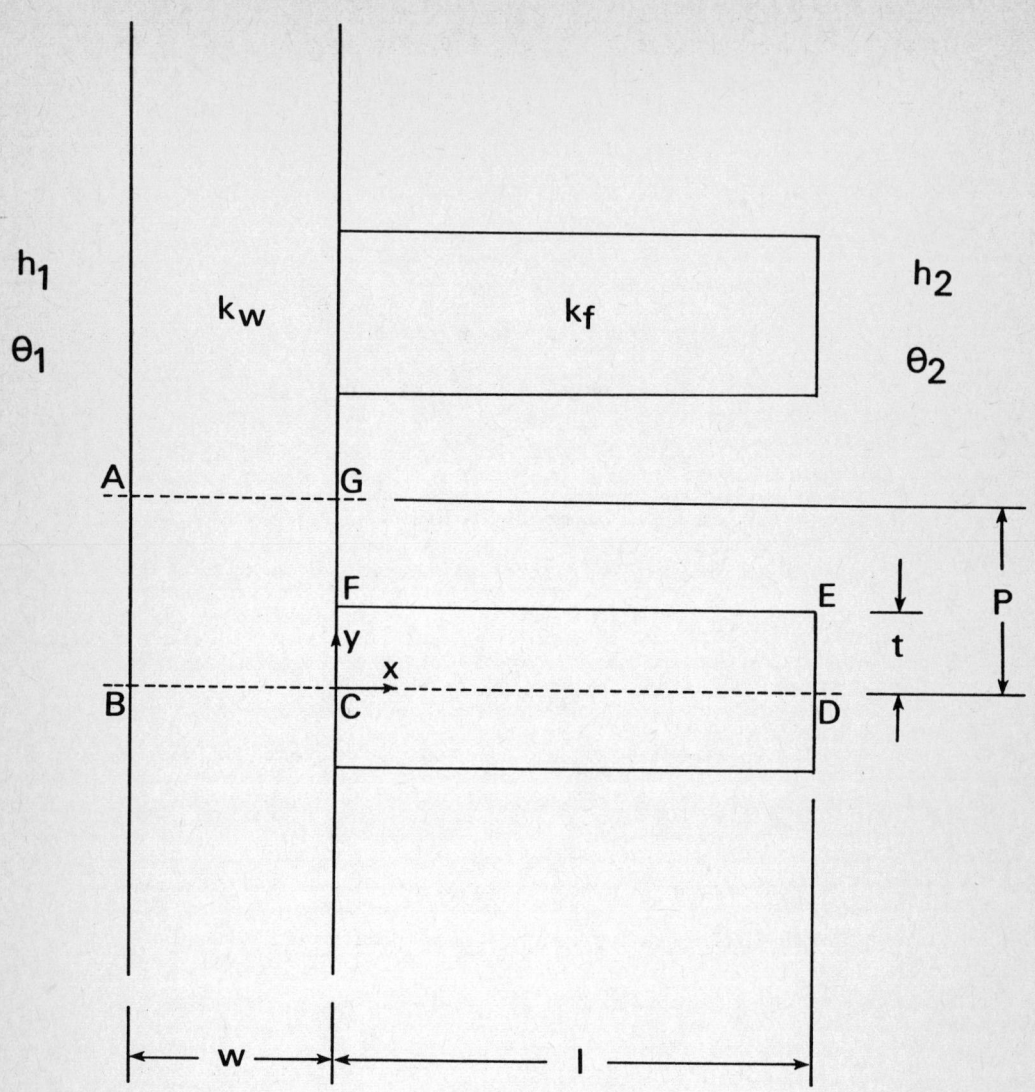

Fig.1 Schematic representation of the fin assembly

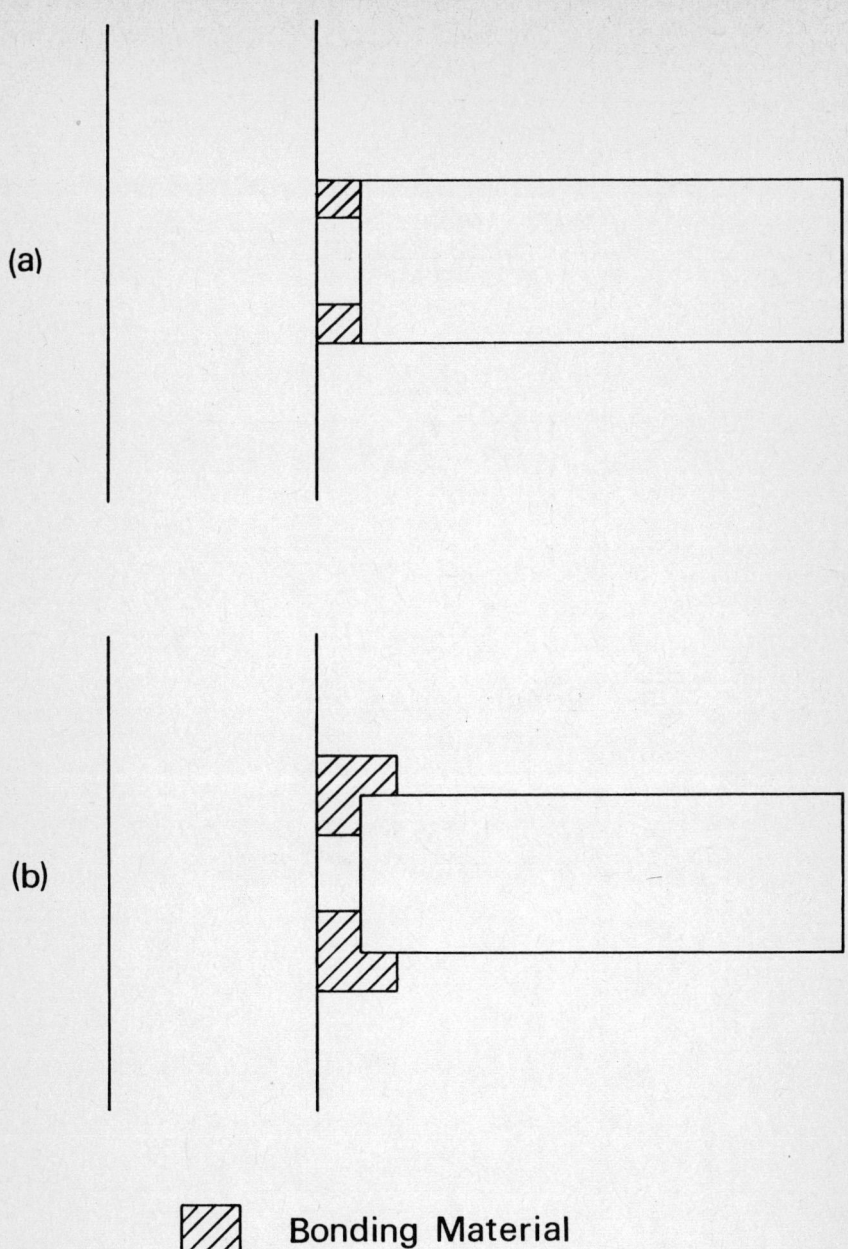

Fig.2 Examples of interfacial bonding

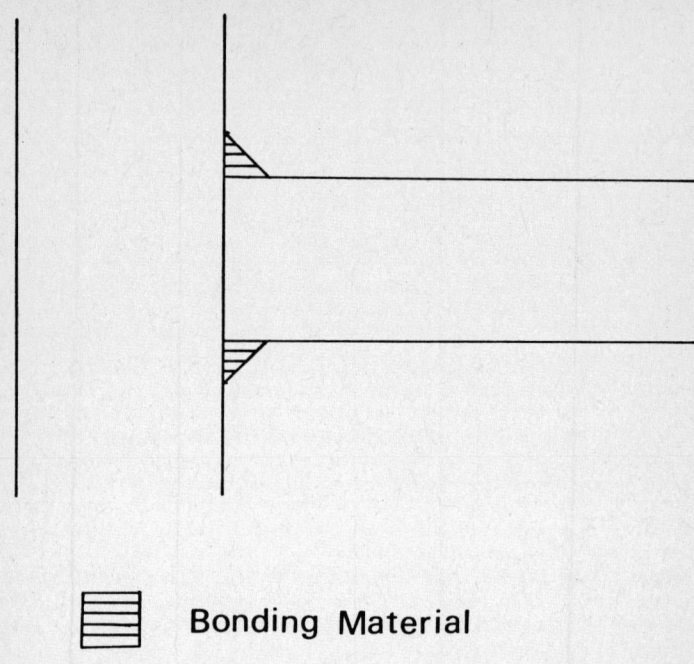

Fig.3 Example of a bond with wall-to-fin contact

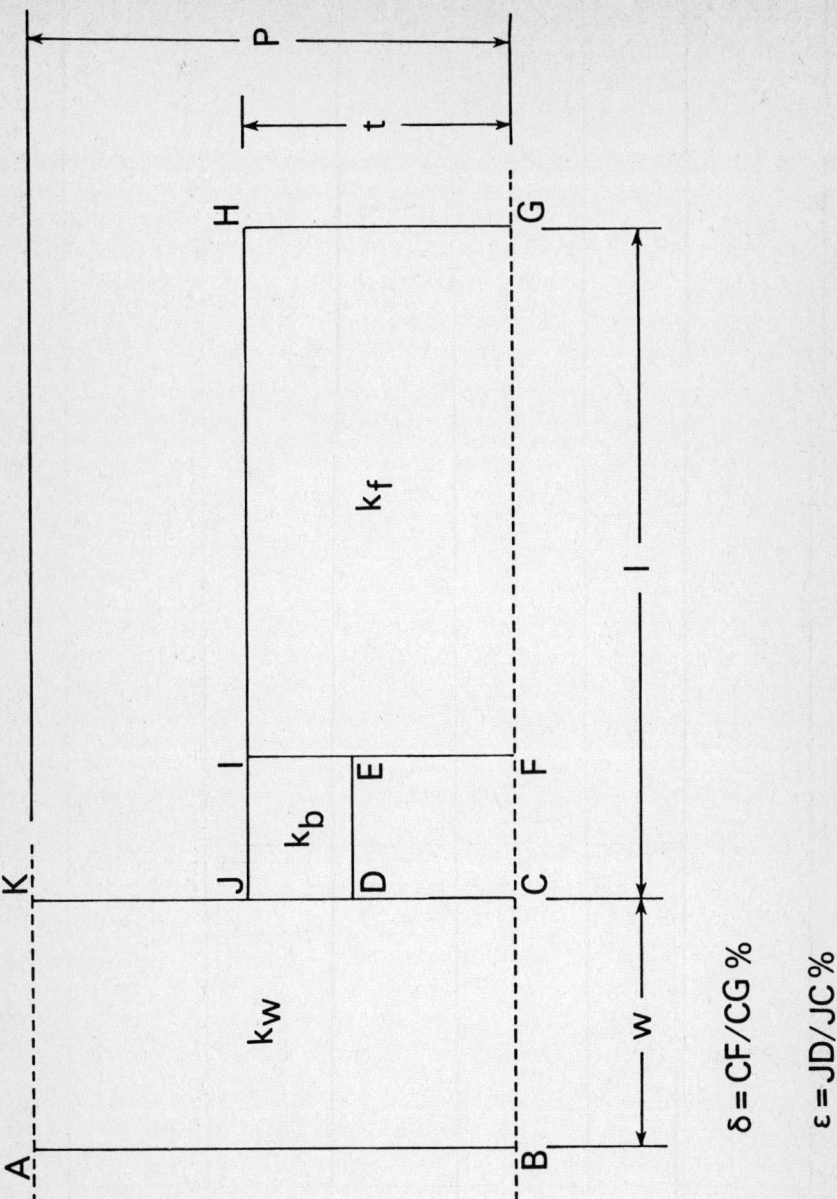

Fig.4 Schematic representation of a fin assembly with interfacial bonding

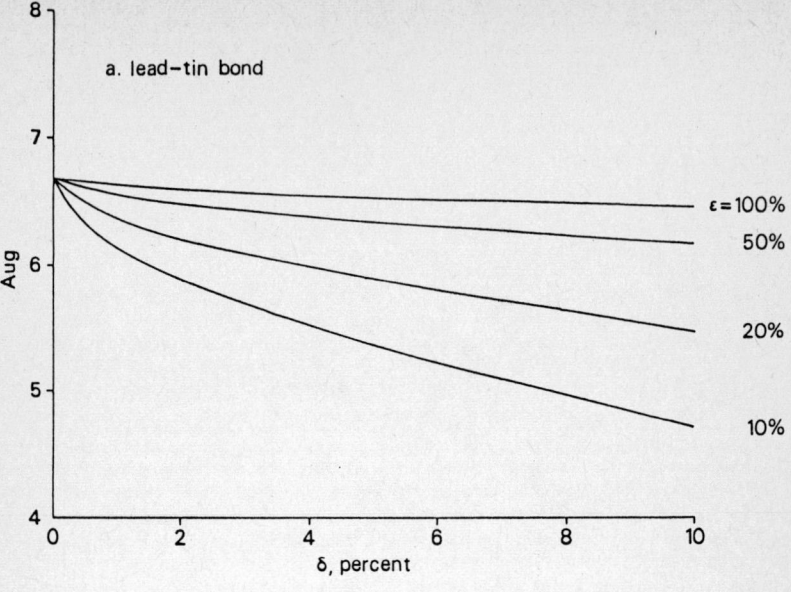

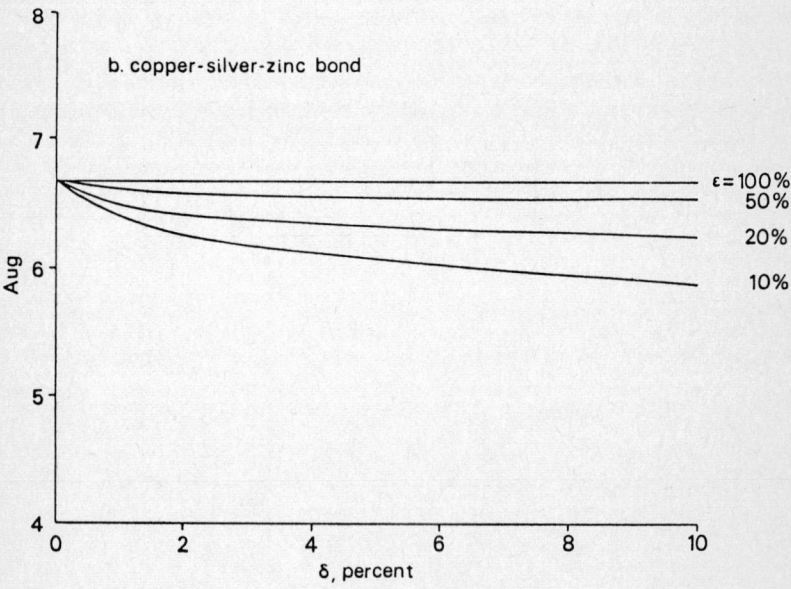

Fig.5 The effects of interfacial bonding for problem A

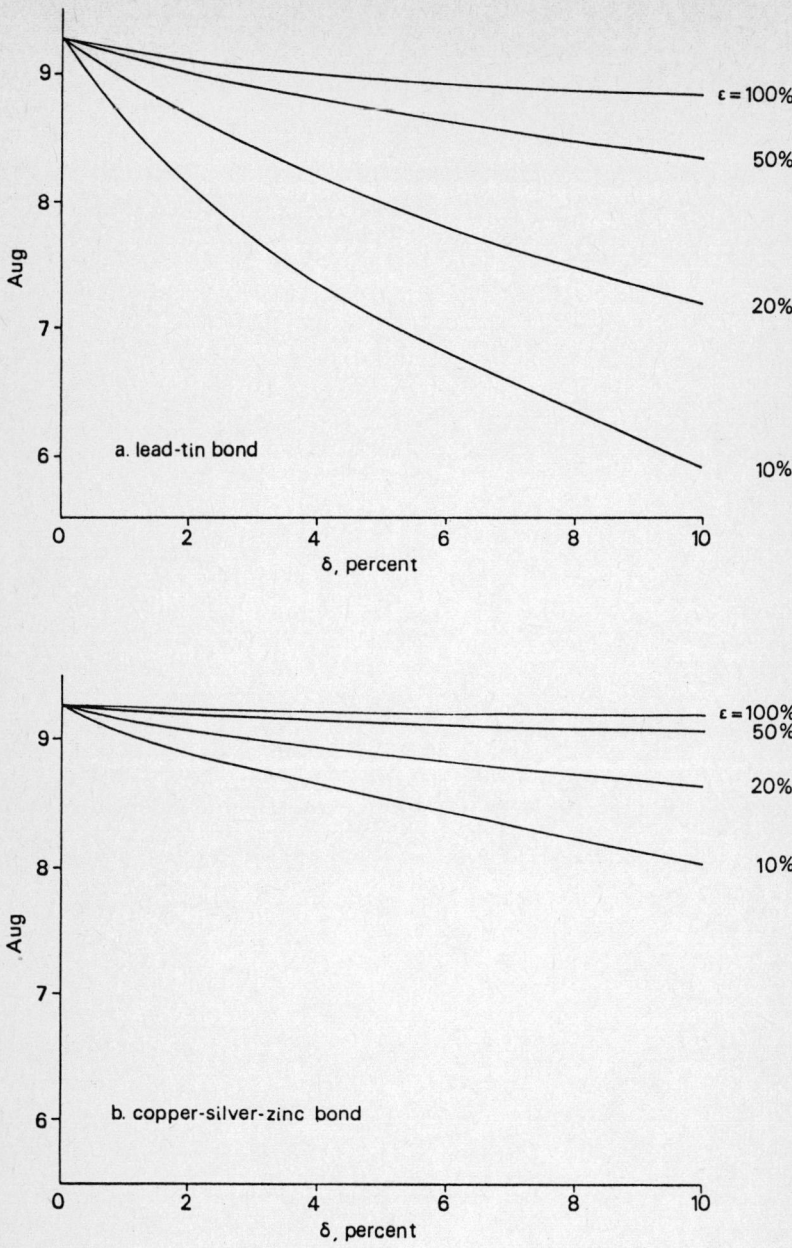

Fig.6 The effects of interfacial bonding for problem B

CHAPTER 6

CONCLUSIONS

6.1 DISCUSSION AND CONCLUSIONS

The work presented in this thesis has been concerned with the investigation of heat transfer problems relevant to the design of extended surface heat exchangers. The main objective has been to establish the applicability of certain simplifying assumptions which are inherent in the present design techniques. This required the examination of considerably more sophisticated mathematical models than previously investigated and involved the development of specialised solution techniques in order to deal with the additional complexities. The most significant feature of the present investigation is that for each problem examined a fin assembly representation has been employed. Thus, the effects of both the thermal interaction between the primary interface and the fins, and the convective heat exchange from the plain side of the primary interface have been explicitly accounted for. In contrast, in the vast majority of the published literature on the subject of extended surface heat transfer attention is restricted solely to the fin side. The concept of considering the primary interface and the fins as separate entities originated from the study performed by Harper and Brown [1] in 1922, and forms the basis of present design techniques. In this approach the fin heat transfer rate is invariably characterised by a dimensionless quantity referred to as the fin efficiency. This fin efficiency is widely considered to be ideal for design purposes because it permits a very concise representation of the fin performance. However, as explained in Chapter 2, this conciseness in fact negates its value for design purposes. The principal reason for this is the fact that fin efficiency charts are usually parameterised by bulk parameters and consequently give little or no indication as to the effects of variations in the individual parameters, such as the fin length, fin thickness, fin thermal conductivity and surface heat transfer coefficient, on the fin performance. Furthermore, the fin efficiency itself gives no indication whatsoever of the overall, i.e. fin assembly, heat transfer rate. This is usually computed by incorporating the fin efficiency into a technique based upon electric circuit theory [2]. In Chapter 2 [3] it was shown that this technique, which is commonly referred to as the Sum of Resistances method, in fact has a mathematically rigorous foundation. It was also shown that if a simple design technique is to be used then there is a far superior alternative to the fin efficiency, namely, the enhancement factor. This enhancement-factor requires more charts in order to describe the fin side heat transfer, but these clearly show the effects of variation in the individual design parameters, and also give some indication as to the fin assembly performance. However, neither the fin efficiency nor the enhancement-factor are suitable for the effective design of finned surfaces because they are both based upon one-dimensional heat flow analyses; a number of recent investigations [4-7] have established that an accurate representation of the heat flow can only be achieved by employing a two-dimensional analysis. These studies emphasised, in particular, that a fin assembly representation must be employed in conjunction with the two-dimensional analyses, because the significant two-dimensional effects result from the thermal interaction between the primary interface and the fins.

The study performed by Stones [7] consolidated the two-dimensional fin assembly approach, but showed that the two most popular solution techniques for handling such problems, namely the finite-difference (FD) and finite-element (FE) methods, do not always yield a convergent solution. The two alternative solution techniques investigated in Chapter 3 [8-12], namely the boundary integral equation (BIE) and series truncation (ST) methods, were found to be far superior. They not only provided convergent solutions in all cases, but also proved to be computationally more efficient than the FD and FE methods. The ST method was found to be the most appropriate solution technique for the case of longitudinal rectangular fins attached to a plane wall in that it gave the most accurate solutions with the minimum computational storage and time requirements. Unfortunately, the ST formulation is limited in applicability to this particular problem. In contrast, the flexibility and conceptual simplicity of the BIE method are such that it can easily handle problems involving curved or tapered fin profiles and non-uniform variations of the surface heat transfer coefficients. Thus, the BIE formulation provides a very powerful solution capability for the analysis of fin assembly heat transfer. In fact, in the context of fin assembly heat transfer, the only signigicant problem encountered with the BIE cormulation has been the inability of the singular BIE method to provide solutions which, in general, converge more rapidly than those predicted by the standard BIE method; significant improvements in accuracy were only achieved for problems involving small fin length to fin thickness ratios. This suggests that for larger ratios, the discretization error is greater than the error associated with the singularity.

The advanced solution capabilities provided by the BIE method are further emphasised by the non-linear BIE formulation, devised in section 4.1 [13], which enables the treatment of problems involving radiative heat dissipation. In fact, of the four aforementioned solution techniques, the BIE method is the only one which permits a practically feasible treatment of the fin assembly problem examined in section 4.2 [14]. The ST method cannot handle the non-linear boundary conditions, and the FD and FE methods are inappropriate because they would require excessive amounts of computational storage and time in order to provide accurate solutions. Furthermore, the classical BIE formulation is perfectly compatible with the analyses which describe the radiant interaction between the various components of the fin assembly. In contrast, the standard FD and FE formulations require tedious modifications similar to those devised for the one-dimensional FD implementation described in section 4.2 [14].

In the linear analysis discussed in Chapter 3 [8-12], the consistency of the solutions predicted by the four different solution techniques was sufficient to validate the accuracy of the computer programs. However, for the non-linear formulation devised in section 4.2 [14], solutions were only computed employing the BIE method. Therefore, in order to establish confidence in the programming, the various facets of radiative heat dissipation were incorporated into the BIE program for the linear problem in simple stages. At each stage the computer program was validated by ensuring that, by assigning appropriate values to the relevant system parameters, the solutions agreed with those predicted at the preceeding stage. For example, at the penultimate stage, the program allowed for black-body radiation and accounted for all relevant radiant interactions; the final program, which of course allowed for gray-body radiation, gave exactly the same results as this program when the surface emissivities of both the primary interface and the fin were prescribed to be unity.

The fin assembly formulation developed in section 4.2 [14] represents a major extension of the previous work on the subject of radiative heat transfer from finned surfaces. This formulation incorporates two very significant features which have not previously been accounted for. Firstly, the effects of the thermal interaction between the primary interface and the fins are included by virtue of the fact that a fin assembly representation is employed, and secondly, the heat flow is accurately modelled by employing a two-dimensional analysis. The results predicted by this formulation confirmed the inadequacy of the previously employed formulations for the effective design of extended surface heat exchangers. However, these results served to emphasise the most significant observation from previous investigations, namely, that the radiative interaction between the various elements on the finned side of the recuperative interface plays a significant role in the heat transfer process except in situations in which the convective heat dissipation is extremely high. Thus, the classical assumption that the surface heat dissipation may be considered to be purely convective is inappropriate for the vast majority of applications.

After the findings of Chapter 4 it would appear futile to perform any analyses which neglect the radiative heat transfer. However, in Chapter 5 an investigation is performed into the applicability of the perfect contact assumption, and in this investigation only convective surface heat transfer is considered. The principal reason for neglecting the radiative heat transfer was to permit the emphasis to be directed towards examining the validity of perfect contact whilst minimising the overall complexity of the problem. Thus, the observations made from the work presented in Chapter 5 [15,16], must be considered to be of a qualitative, rather than quantitative, nature.

In the published literature there appears to be only three articles on the subject of reduced wall-to-fin contact [15,16,17]. All these articles have been concerned with problems in which the contact is completely relaxed. However, in most applications some form of bonding is used in order to maintain contact, at least at discrete zones along the contact interface. Thus, the models of surface roughness and interfacial bonding developed in Chapter 5 [18,19] should facilitate an improvement in the design of extended surface heat exchangers because they are more representative of the types of contacts encountered in practice. Furthermore, from a practical viewpoint, the models investigated in this study are even superior to the most sophisticated models examined in the vast range of published material on the subject of thermal resistance between contacting surfaces, e.g. [20-24]. This is primarily due to the fact that in this study the emphasis has been directed towards examining the effects of reduced contact on the overall heat flow by employing relatively simple models of the contact interface. In contrast, in the previous publications the emphasis has been directed towards actually modelling the complex features which govern contact.

The conceptual simplicity of the contact interface models investigated in this study is not indicative of the considerable complexity which these models present for a mathematical treatment. In fact, only the flexibility of the BIE discretization is capable of handling these models without difficulty. With the ST method the analysis becomes very laborious, in particular, for the model of interfacial bonding, whilst with the FD and FE methods the tedious nature of the discretization prcoesses virtually precludes treatment. Fortunately, the results predicted by these models indicate that it is unnecessary to consider more sophisticated representations of the contact interface. In fact, the results showed that the perfect contact assumption will suffice for the types of contacts encountered in practice.

Thus, to summarise, the present investigation has established, in particular, that

i   an accurate representation of the heat flow within a recuperative interface can only be achieved by examining a fin assembly model in conjunction with a multi-dimensional analysis,

ii the radiative heat dissipation plays a significant role in the heat transfer process and therefore should not be neglected, and

iii the assumption that there is perfect contact between the primary interface and the fins may be employed provided contact can be achieved over at least 10 per cent of the contact interface.

In view of the conclusions of this study, it is suggested that future research should be directed towards developing analyses which overcome the need to use uniform heat transfer coefficients. The inapplicability of this simplification is intuitively obvious, but has been further emphasised by the findings of numerous experimental investigations [7, 25-28]. In particular, Stones [7] has shown that, for the case of annular fins attached to a cylindrical tube, the use of averaged uniform heat transfer coefficients can result in errors of over 400 per cent in the prediction of the heat transfer rate. The simplest method of introducing non-uniformity is to employ assumed variations of the heat transfer coefficients, [29,30]. However, in general, it is not possible to determine which variation to apply for a specific problem. Thus, it would appear that the only alternative is to actually model the convective heat transfer more accurately. This can be achieved either by examining the complicated inter-relation between heat and mass transfer, or by employing certain empirical relations. The mathematical complexity of the former is such that treatment would first require the formulation of an adequate theoretical representation and the development of appropriate solution techniques. However, the latter seems practically more feasible. In fact, such an approach has been attempted in a recent paper by Karvinen [31]. Although the analysis in this paper was performed on the basis of one-dimensional heat flow with attention restricted solely to the fin side, the relations used to describe the surface convection appear to be perfectly compatible with the non-linear BIE method devised in section 4.1 [13].

REFERENCES

1. D.R. Harper and W.B. Brown, "Mathematical equations for heat conduction in the fins of air-cooled engines", National Advisory Committee for Aeronautics, Report 158, 1922.

2. F. Kreith, Principles of Heat Transfer, Harper and Row, New York, 1976.

3. P.J. Heggs, D.B. Ingham and M. Manzoor, "The one-dimensional analysis of fin assembly heat transfer", submitted to Journal of Heat Transfer, 1981.

4. E.M. Sparrow and L. Lee, "Effects of fin-base temperature depression in a multifin array", Journal of Heat Transfer, Vol. 97, pp. 463-465, 1975.

5. N.V. Suryanarayana, "Two-dimensional effects on heat transfer from an array of straight fins", Journal of Heat Transfer, Vol. 99, pp. 129-132, 1977.

6. P.J. Heggs and P.R. Stones, "The effects of dimensions on the heat flowrate through extended surfaces", Journal of Heat Transfer, Vol. 102, pp. 180-182, 1980.

7. P.R. Stones, Ph.D. Thesis, University of Leeds, 1980.

8. D.B. Ingham, P.J. Heggs and M. Manzoor, "The numerical solution of plane potential problems by improved boundary integral equation methods", Journal of Computational Physics, Vol. 42, pp. 77-95. 1981.

9. D.B. Ingham, P.J. Heggs and M. Manzoor, "Boundary integral equation analysis of transmission line singularities", IEEE Transactions on Microwave Theory and Techniques, Vol. 29, pp. 1240-1243, 1981.

10. P.J. Heggs, D.B. Ingham and M. Manzoor, "Boundary integral equation analysis of fin assembly heat transfer", to appear in Numerical Heat Transfer.

11. P.J. Heggs, D.B. Ingham and M. Manzoor, "The analysis of fin assembly heat transfer by a series truncation method", to appear in Journal of Heat Transfer.

12. D.B. Ingham, P.J. Heggs and M. Manzoor, "The two-dimensional analysis of fin assembly heat transfer: A comparison of solution techniques", to appear in the Proceedings of the Second National Symposium on Numerical Methods in Heat Transfer, Hemisphere, Washington, D.C., 1982.

13. D.B. Ingham, P.J. Heggs and M. Manzoor, "The boundary integral equation solution of non-linear plane potential problems", to appear in the Institute of Mathematics and Its Applications Journal of Numerical Analysis.

14. D.B. Ingham, P.J. Heggs and M. Manzoor, "Improved formulations for the analysis of convecting and radiating finned surfaces", to appear in AIAA Journal.

15. K.A. Gardner and T.C. Carnovos, "Thermal-contact resistance in finned tubing", Journal of Heat Transfer, Vol. 82, pp. 279-284, 1960.

16. E.H. Young and D.E. Briggs, "Bond resistance of bimettalic finned tubes", Chemical Engineering Progress, Vol. 61, pp. 71-76, 1965.

17. M.V. Kulkarni and E.H. Young, "Bitmetallic finned tubes", Chemical Engineering Progress, Vol. 62, pp. 69-74, 1966.

18. P.J. Heggs, D.B. Ingham and M. Manzoor, "The effects of surface roughness on the performance of extended surface heat exchangers", submitted to Journal of Heat Transfer, 1981.

19. P.J. Heggs, D.B. Ingham and M. Manzoor, "The effects of interfacial bonding on the performance of extended surface heat exchangers", submitted to Journal of Engineering for Industry, 1981.

20. K. Sanokawa, "Heat transfer between metallic surfaces in contact", Bulletin of the Japanese Society of Mechanical Engineers, Vol. 11, pp. 253-263, 1968.

21. J.R. Barber, "The effect of thermal distortion on constriction resistance", International Journal of Heat and Mass Transfer, Vol. 14, pp. 751-766, 1971.

22. B.B. Mikic, "Thermal contact conductance: Theoretical considerations", International Journal of Heat and Mass Transfer, Vol. 17, pp. 205-214, 1974.

23. J.R. Howard, "Heat transfer between contacting solids", Journal of Engineering, Vol. 215, pp. 220-222, 1975.

24. A.M. Jones, P.W. O'Callaghan and S.D. Probert, "Thermal rectification due to distortions induced by heat fluxes across contacts between smooth surfaces", Journal of Mechanical Engineering Science, Vol. 17, pp. 252-261, 1975.

25. P.W. Wong, "Mass and heat transfer from circular finned cylinders", Journal of the Institution of Heating and Ventilating Engineers, Vol. 23, pp. 1-23, 1963.

26. J.W. Stachiewicz, "Effect of variation of local film coefficient on the fin performance", Journal of Heat Transfer, Vol. 91, pp. 21-26, 1969.

27. V.F. Yudin and L.F. Tokhtorora, "Investigation of the correction factor ψ for the theoretical effectiveness of a round fin", Thermal Engineering, Vol. 20, pp. 66-78, 1973.

28. P.J. Heggs and P.R. Stones, "Improved design methods for finned tube heat exchangers", Transactions of the Institution of Chemical Engineers, Vol. 58, pp. 147-154, 1980.

29. L.S. Han and S.G. Lefkowitz, A.S.M.E. Paper 60-WA-41, 1960.

30. P.J. Heggs, D.B. Ingham and M. Manzoor, "The effects of non-uniform heat transfer from an annular fin of triangular profile", Journal of Heat Transfer, Vol. 103, pp. 184-185, 1981.

31. R. Karvinen, "Natural and forced convection heat transfer from a plate fin", International Journal of Heat and Mass Transfer, Vol. 24, pp. 881-885, 1981.

B